The Open University

Science: A Second Level Course

BIOCHEMISTRY

5 – 6   Regulation of Cell Processes

Prepared by an Open University Course Team

THE OPEN UNIVERSITY PRESS

THE BIOCHEMISTRY COURSE TEAM

Steven Rose (*Chairman and General Editor*)
Norman Cohen
Jeff Haywood
Brian Tiplady
Eve Braley-Smith (*Editor*)
Robin Harding (*Course Assistant*)
Bob Cordell (*Staff Tutor*)
Vic Finlayson (*Staff Tutor*)
Roger Jones (*BBC*)
Jim Stevenson (*BBC*)

The Open University Press
Walton Hall, Bletchley, Buckinghamshire

First published 1972

Copyright © 1972 The Open University

All rights reserved.
No part of this work may be reproduced in any form, by mimeograph or any other means, without permission in writing from the publishers.

Designed by the Media Development Group of the Open University.

Printed in Great Britain by
Staples Printers Limited
at St Albans

SBN 335 02052 6

This text is one in a series of units that make up the correspondence element of an Open University Second Level Course. The complete list of units in the course is given at the end of this text.

For general availability of supporting material referred to in this text, please write to the Director of Marketing, The Open University, Walton Hall, Bletchley, Buckinghamshire.

Further information on Open University courses may be obtained from the Admissions Office, The Open University, P.O. Box 48, Bletchley, Buckinghamshire.

UNITS 5–6   REGULATION OF CELL PROCESSES

Contents

Table A	List of Scientific Terms, Concepts and Principles	4
Objectives for Units 5–6		6
Study Guide to Units 5–6		7
Pre-Unit Assessment Test		8

UNIT 5

5.1	Introduction	9
5.2	The Regulation of Enzyme Activity	11
5.2.1	Passive stability	11
5.2.2	Techniques for studying the sequences of metabolic pathways	12
5.2.3	Feedback inhibition in a linear pathway	15
5.2.4	Feedback inhibition in branched pathways	17
5.2.5	Evolution of feedback systems	20
	SAQ for Sections 5.2.3–5.2.4	21
5.3	Allosteric Inhibition	22
5.4	Cell Compartmentation	22
	SAQs for Sections 5.3–5.4	23
5.5	Summary of Unit 5	23

UNIT 6

6.1	Introduction	24
6.2	Synthesis and Degradation of Proteins	26
	SAQs for Sections 6.1 and 6.2	27
6.3	The Mechanism of Protein Synthesis	28
	SAQs for Section 6.3	29
6.4	The Mechanism of Induction and Repression in Bacteria	30
6.4.1	The lactose system	31
6.4.2	Summary of Section 6.4	34
	SAQs for Section 6.4	35
6.5	The Regulation of Protein Synthesis in Higher Organisms	35
6.5.1	A critical look at the applicability of the Jacob-Monod hypothesis to higher organisms	36
	The induction of tryptophan pyrrolase (TP)	37
6.6	The Control of DNA Synthesis and Cell Division	40
6.7	Conclusions to Units 5–6	41
	Conclusion to the Course	43
	Answers to Pre-Unit Assessment Test	44
	Answers to In-text Questions	44
	Self-assessment Answers and Comments	46

Table A
List of Scientific Terms, Concepts and Principles used in Units 5–6

Taken as prerequisites			Introduced in this Unit	
1 Assumed from general knowledge	2 Introduced in a previous Unit	Unit No.	3 Developed in this Unit or in its set book	Page No.
		S100*		In Unit
bacterium	half-life	2	genetic continuity	9
fungus	isotope	6	model cell	12
growth	radioactivity	6	biochemical vat	12
organ	ion	9	passive stability	12
reproduction	pH	9	isotope competition	13
tissue	amino acid	10	metabolic mutants	14
	glucose	10	feedback inhibitor	16
	polysaccharide	10	end-product inhibitor	16
	protein	10	sensitive enzyme	16
	rate of reaction	11	allosteric	17
	catalysis	12	enzyme multiplicity	17
	dynamic equilibrium	12	multiproduct inhibition	19
	macromolecule	13	sequential feedback	20
	monosaccharide	13	end-product repression	24
	peptide bond	13	enzyme repression	24
	ATP	14	enzyme induction	25
	cell membrane	14	inducer	26
	mitochondrion	14	inducible	26
	nucleus	14	repressible	26
	ribosome	14	pro-enzyme	30
	cofactor	15	constitutive mutant	32
	competitive inhibitor	15	inducibility	32
	enzyme	15	regulator gene	32
	glycolysis	15	structural gene	32
	metabolic pathway	15	operon	32
	NAD	15	repressor	32
	substrate	15	promoter	33
	virus	15	operator region	33
	cell compartmentation	16	initiation of DNA replication	40
	cybernetics	16	chain growth in DNA replication	40
	end-product inhibition	16 (TV)		
	feedback control	16	termination of DNA replication	40
	feedback inhibition	16		
	regulatory mechanisms	16		
	activating enzyme	17	polynucleotide	40

* *The Open University* (*1970*) *S100* Science: A Foundation Course, *The Open University Press*.

Table A continued

Taken as prerequisites			Introduced in this Unit	
1 Assumed from general knowledge	2 Introduced in a previous Unit	Unit No.	3 Developed in this Unit or in its set book	Page No.
	cell division	17		
	codon	17		**In**
	DNA	17		**CS & F****
	DNA replication	17	allosteric sites	259
	double helix	17	cooperativity	263
	gene	17	microsome	378
	genetic code	17	polysome	380
	mRNA	17	initiator codon	394
	nucleic acid	17	terminator codon	395
	nucleotide	17		
	phage	17		
	RNA	17		
	RNA polymerase	17		
	transcription	17		
	tRNA	17		
	translation	17		
	homeostasis	18		
	hormone	18		
	physiology	18		
	teleology	18		
	thyroxine	18		
	mutant	19		
	evolution	21		
		S2–1		
	multienzyme complex	1		
	subunit	1		
	specific activity	1		
	isoenzyme	2		
	quasi-substrate	2		
	specific radioactivity	2		
	activation	4		
	citric acid cycle	4		
	pentose phosphate pathway	4		
	starting metabolite	4		

** *A. G. Loewy and P. Siekevitz (1969)* Cell Structure and Function, *2nd ed., Holt, Rinehart and Winston.*

Objectives for Units 5–6

By the end of these Units you should be able to:

1 Understand all the terms, concepts and principles given in Table A. Demonstrate this understanding by:
(a) distinguishing between the terms, concepts and principles;
(b) distinguishing between true and false statements about them.
(SAQs 2, 4.)

2 List the major means of regulating cell metabolism.

3 State two techniques for studying metabolic pathways.
(ITQ 3, SAQ 5.)

4 Predict the consequences of a change in level of one or more components of a given metabolic pathway, given the regulatory controls of the pathway.
(ITQ 2, SAQ 1.)

5 Distinguish between valid and invalid hypotheses concerning the regulation of cell metabolism, given appropriate data.
(ITQ 3, SAQs 2, 3, 4, 5, 9, 10.)

6 Draw a diagram to represent the overall transfer of information from DNA to protein. Understand the mechanism of this transfer.
(SAQs, 7, 8.)

7 List (or select from a list) the components necessary for synthesis of protein in a cell-free system.
(ITQ 3, SAQ 6.)

8 Represent in diagrammatic form (or label a given diagram of) the Jacob-Monod hypothesis on induction and repression.
(ITQ 4.)

9 State the main features of the Jacob-Monod hypothesis and indicate this knowledge by evaluating experiments and systems purporting to support or refute the hypothesis.
(ITQs 3 and 4, SAQs, 9, 10.)

10 Describe the three main steps in the synthesis of DNA in bacteria.

Study Guide to Units 5–6

These two Units cover one topic, the regulation of cellular processes. The division into two Units is arbitrary. Both Units assume a knowledge of S100, particularly of Units 16 and 17. You should therefore attempt the Pre-Unit Assessment Test given on p. 8. You should check your answers against those given on p. 44 before proceeding with the text of these Units. At the end of some Sections of the text you will find further Self-assessment Questions (*SAQ*s). We feel that you will benefit most by doing them as they arise and checking your answers against those given before continuing with the next Section of the text.

You are not required to remember chemical formulae or names of chemicals or enzymes mentioned in the text unless specifically directed to do so. You may, however, find that remembering some illustrative examples assists you in remembering the principles that they illustrate, as required in Objectives 1 to 10. The text may seem rather brief. Do not, however, be misled! In addition to the text you are directed to read 34 pages of *Cell Structure and Function* – 9 pages in Section 5.3 and 25 pages in Section 6.3. The latter reading (6.3), however, is very much a recapitulation of parts of S100, Unit 17.

These Units are really one Unit, so you can usefully break your studies at any one of several points. We think the division we have chosen between Units 5–6 is a convenient place to break off. However, the end of Section 6.3 is almost as convenient, should you find that you complete Unit 5 quickly.

We do not feel that any Sections can really be omitted; they are all of equal importance. However, *if* you are very short of time, there are ways in which you could save some time:

(a) Section 6.3 could be replaced by a careful reading of the relevant portions of Unit 17 of S100.

(b) You could read the comments in the structured exercise in Section 6.5.1 immediately after reading each block of data and interpretations.

(c) You *could* omit Section 6.6, thus neglecting only Objective 10.

We hope, of course, that you find these measures unnecessary.

One further point. Throughout these Units you will find reference to studies on one organism, the bacterium *Escherichia coli*. The reasons for this are simple. As you will see from the TV programme for Unit 5, bacteria are easy to handle and *E. coli* is no exception. This has made it a very popular laboratory organism. Given a frugal medium comprising NH_4^+ ions as a source of nitrogen, K^+, Mg^{2+}, Na^+, SO_4^{2-} ions as a source of sulphur, PO_4^{3-} as a source of phosphorus, small amounts of other metal ions and *any one* of twenty or so organic compounds (e.g. glucose, acetic acid) as a source of carbon, *E. coli* will grow and divide rapidly. Indeed, perhaps we should dedicate these Units to *E. coli*, an organism without peers as a contributor to biochemistry and molecular biology!

Pre-Unit Assessment Test

Study Comment

You should now attempt this test, then check your answers against those on p. 44, before proceeding with these Units.

Section A

Questions 1 and 2 In the following reaction sequence:

L → M → N → O

L-ase, M-ase, and N-ase are enzymes. The maximum rate of reaction catalysed by L-ase is 100 units/min, by M-ase 15 units/min, by N-ase 25 units/min.

1 Which is the rate-limiting step in the above sequence?

2 If O inhibits L-ase, this is an example of:
    A competitive inhibition
    B repression
    C catalysis
    D feedback inhibition

Choose the correct alternative.

Questions 3–7 Indicate which of the following statements are true and which are false:

3 Cell membranes are capable of controlling the rate of uptake or excretion of various chemicals.

4 The rate of ATP utilization tends to affect its rate of synthesis.

5 Feedback control only operates at the subcellular level.

6 Hormones are substances that are important for homeostasis in multicellular organisms.

7 Cellular control mechanisms are unlike any utilized by engineers in constructing machinery.

Section B

Questions 1–7 Indicate which of the following statements are true and which are false:

1 Both DNA and RNA contain the sugar ribose.

2 The two strands of the DNA double helix are held together by hydrogen bonds.

3 In the genetic code, for each amino acid there is only one codon.

4 For each mRNA there is one specific amino acid.

5 Protein synthesis occurs on ribosomes.

6 Ribosomes are composed of two subunits.

7 Transcription occurs on ribosomes.

Section C

Questions 1–5 Which of the following statements are teleological?

1 Birds have wings in order to fly.

2 Cells contain certain enzymes so that they can hydrolyse proteins.

3 The role of the mitochondrion is to provide ATP for the cell.

4 Lipids form part of the structure of cell membranes.

5 Cell membranes exclude certain chemicals from the cell.

6 The nature of mitochondrial function is such that it provides ATP for the cell.

When you have finished all three sections (A–C) of this test, turn to the answers on p. 44, before proceeding with these Units.

UNIT 5

5.1 Introduction

Many bacteria are capable of growth and reproduction under a wide variety of environmental conditions. For example, one such bacterium, *Alcaligenes faecalis*, can grow in media varying in pH from pH 5 to pH 9. The common intestinal bacterium, *E. coli*, can grow at temperatures between 15 °C and 42 °C, in highly viscous glycerol-containing media, and can utilize a wide variety of 'foods' – over twenty different compounds can each serve as sole source of carbon. All these variations affect the rate of growth and reproduction, but these processes do still occur.

If these statements strike you as unremarkable, then consider what growth and reproduction of bacteria entail. To grow and reproduce itself, a bacterial cell must double its own content of macromolecules (nucleic acids, polysaccharides, proteins). To do this, the bacterium must convert the small molecules which it obtains from the surrounding medium to provide both the building-blocks (nucleotides, monosaccharides, amino acids) and the energy (ATP) for all the syntheses. Having doubled in size, the bacterial cell then divides to give two daughter cells. However, to ensure that the two daughter cells are identical, and hence maintain genetic continuity (that is, the passage of hereditary information from generation to generation), this division must be preceded by accurate replication of the DNA, followed by exact apportioning of the DNA into each of the two parts of the cell destined to become the two new cells. Thus each daughter cell inherits the same complement of genetic information (S100, Unit 17). This process necessitates a link in timing between DNA synthesis and cell division.

growth and reproduction

As you know from the preceding Units of this Course, the vast numbers of different reactions that occur within a cell are facilitated by enzymes. But enzymes are very sensitive to changes in pH, temperature, concentrations of small molecules, etc. Furthermore, the activity of each enzyme varies in a different way from that of any other enzyme in response to these parameters. So if any of these parameters (pH, temperature, etc.) were altered in the medium surrounding a bacterial cell, one would expect the catalytic rates of different enzymes to respond differently. Indeed, changes in such parameters do lead to changes in the rate of growth and reproduction of bacteria. But the very fact that the cells still grow and reproduce at all argues for their ability to adjust to the new conditions and thus maintain enough overall stability to regulate growth and reproduction, despite the differential effects on different enzymes. Thus, cells must be able to regulate overall processes, like growth and reproduction, by controlling the relative rates of individual reactions in these processes. This control is necessary for adjusting to changes in environmental conditions. It is also necessary for cellular economy: the cell gears its production of various compounds to the demand for them in other processes. In these two Units (5–6), we will discuss what is known and, perhaps more important and certainly more extensive, what is not known about regulatory mechanisms in cells.

We introduced this Unit by referring to bacteria. This was not fortuitous. Much of what is known about regulatory mechanisms has been learnt from the study of bacteria. This is particularly true for the regulation of protein synthesis, a topic we discuss in Unit 6. This does not, of course, mean that regulatory mechanisms exist only in bacteria. Indeed, in Unit 4 the regulatory systems described were ones studied in higher organisms, such as mammals. However, the relative ease of handling (TV programme of Unit 5), the ability of the investigator to easily change the environment surrounding the cells, and the availability of genetic variants (mutants) have all made bacteria popular experimental organisms. The practical advantages of using bacteria and their viruses (phages) was appreciated in the 1930s and 1940s, notably by Max Delbruck, a physicist working in the United States, and André Lwoff and Jacques Monod, two biologists working in Paris. This appreciation, coupled to the development of techniques of genetic analysis in bacteria by Joshua Lederberg and others, led to an explosion in our understanding of cellular control mechanisms—a large branch of the science called molecular biology.*

Still, one cannot justifiably ignore all organisms other than bacteria, and we shall certainly not do so in these Units. We shall, however, tend to approach the subject from what is known in bacteria and pose the question: how relevant, or otherwise, are such findings to multicellular organisms? This approach is very prevalent in current research in the field of regulatory mechanisms. It has, however, left the molecular biologist open to the jibe that he believes that 'what is true for *E. coli* is also true for *E. lephant*'. Perhaps, if one considers each cell in an elephant as an entity in itself and all the other cells as part of that cell's external environment, the subcellular control mechanisms in each cell might be basically like those in *E. coli*. Even if true, this would still ignore the need the elephant has to regulate itself as a whole. There, one must consider interactions between cells, between tissues, between organs and systems of organs. Some of these multicellular interactions have been considered in S100, Unit 18, and in Unit 4 of this Course, and are broadly speaking part of physiology. In these two Units, we shall restrict our discussion to subcellular regulation, only considering regulation between cells where it is likely to effect the basic subcellular mechanisms. Nevertheless, you should keep in mind the existence of these various levels of control in complex organisms. They are particularly relevant when considering development of organisms, and are dealt with in the Course *Genetics and Development* (S2–5) (starting in 1973).

From the above introductory remarks, you may have realized that one of the problems inherent in investigating regulatory mechanisms, or indeed in writing Course Units about them, is, 'where does one begin?' By their very nature, regulatory mechanisms are 'circular'—break the circle and the regulation breaks down. However, as we have already argued that enzymes are vital to cellular activity and are responsive to changes in conditions, it seems reasonable to suppose that some regulatory mechanisms might operate by controlling the rates of enzyme-catalysed reactions. As you will recall from Unit 2, the rate of an enzyme-catalysed reaction basically depends on two features: the amount of enzyme present (i.e. double the enzyme, double the rate of reaction); and the composition of the medium surrounding the enzyme. So, in principle, the rate of an enzyme-catalysed reaction can be controlled in two ways:

> 1 By regulating the rate at which a fixed amount of the enzyme catalysed the reaction, depending on the composition of the medium inside the cell.
>
> 2 By regulating the amount of the enzyme in the cell.

In Unit 4, you considered some specific examples of the control of certain metabolic pathways by mechanism 1. In Unit 5, we shall discuss the theoretical aspects of this mechanism, without leaning heavily on specific examples. In Unit 6, we shall discuss control via mechanism 2.

One further point is worth noting. You will already perhaps have noticed a degree of teleology creeping into these Units. You have been told about the dangers of teleology in S100, Unit 18, and further told that teleological arguments may be seen merely as shorthand versions of non-teleological statements. We use teleological arguments in these Units as such a shorthand.

teleology

But we also use them in another way. In a 'circular' field, such as that of regulatory mechanisms, often the only way to initiate a scientific investigation or discussion is to put oneself, as it were, in the place of the cell—how would I do such and such, given . . . ? The temptation to do this is overwhelming, particularly since some workers in this area have borrowed the terminology of cybernetics and the way of thinking of control engineers and computer programmers. This teleological way of thinking can be useful and so we, the Course Team, make no apology for adopting it in these two Units. We do, however, offer this warning: be able to recognize which arguments are teleological and appreciate that we are certainly *not* suggesting that the cell knows what it wants!

---

* Those of you interested in the history of molecular biology can read a fascinating account containing much amusing anecdotal material in J. Cairns, G. S. Stent and J. D. Watson (ed.) (1966) Phage and the Origins of Molecular Biology, Cold Spring Harbor Laboratory of Quantitative Biology.

5.2 The Regulation of Enzyme Activity

Study comment

In this Section we discuss the regulation of metabolic pathways by means of controlling the activity of particular enzymes in those pathways. You need not remember any of the specific examples given but, to achieve Objective 4, you should understand the basic features of the control systems operating in linear and branched pathways.

One way of regulating the overall through-put of a metabolic pathway is to control the activity of one or more enzymes in that pathway (S100, Unit 16; Unit 4 of this Course). In discussing such regulation, it is convenient to arbitrarily divide enzymes into two broad classes:

I Enzymes involved in intermediary metabolism, i.e. catalysing reactions whereby small molecules are converted to other small molecules. This class encompasses all those enzymes involved in pathways wherein substances are broken down and ATP and $NADH_2$ are produced (Units 3 and 4), and in those pathways leading to the production of the building-blocks (amino acids, monosaccharides, nucleotides) for the synthesis of macromolecules.

II Enzymes catalysing reactions in which building-blocks are linked together to form macromolecules, e.g. RNA polymerase—the enzyme which catalyses the linking of nucleotides to form RNA.

In this Section, we shall confine our discussion to the enzymes in class I, though some of the principles we deal with also apply to those in class II.

5.2.1 Passive stability

It is relatively easy to study the control of the activity of a single enzyme. The enzyme can be isolated, purified and its specific activity measured under a variety of conditions. Various compounds known to be involved in the pathway in which the enzyme occurs, can be added to the pure enzyme and their effects on the rate of reaction noted. In this way, one can build up a picture of how the activity of the enzyme *might* be controlled and how this *might* help regulate the pathway in which it occurs, *in vivo*. The operative word is 'might'. There are several problems. As was mentioned in Unit 4, the investigator tends to examine the effect of compounds that *he expects to be relevant to the pathway*. Furthermore, even if a particular compound connected with the pathway in question is found to affect the activity of an enzyme of the pathway *in vitro*, it is still necessary to establish the importance of this effect *in vivo*. For example, the inhibition of phosphofructokinase (PFK) by ATP and citric acid and its activation by ADP and AMP (Unit 4) *in vitro*, does not prove a central role for this enzyme in the regulation of glycolysis *in vivo*, though it suggests such a role. One must first establish that the levels of the inhibitors or activators that occur *in vivo* are actually high enough to produce the effects on the enzyme that are observed *in vitro*. Then, account must be taken of other possible control points in the pathway besides the one investigated. As you saw from Unit 4, many such points can exist in a single pathway; carbohydrate metabolism seems to have several, including hexokinase, phosphorylase and PFK. Even if all the enzymes of a pathway are purified and their control by various compounds elucidated *in vitro*, it is still a major task to evaluate the relative contributions of the individual enzymes to the regulation of the pathway as a whole, *in vivo*. The number of variables may be very great and often computers are needed to handle them. Nevertheless, it is usually possible from data on individual enzymes to construct some hypothesis of how the regulation of a whole pathway operates. One test of such a hypothesis is to programme a computer with all the experimental data plus the hypothetical regulation scheme and to instruct the computer to behave like the real pathway, that is, to predict the behaviour of the pathway under a variety of conditions. If the predicted behaviour accurately mimics the observed behaviour of the whole pathway *in vivo*, then one can feel reasonably confident that the data are accurate and the hypothesis tenable. Several attempts have been made at such computer-simulated pathways—glycolysis in yeast and glucose metabolism in brain, to name but two. The fit between the computer predictions and experimental observations on the real pathways have been quite good but not perfect. This indicates that not all the mechanisms at work in these pathways have been uncovered.

Imagine magnifying this problem to the extent of trying to predict the behaviour of a whole cell on the basis of the detailed knowledge of its individual reactions, pathways and their interconnections.

One can, however, simplify the concept of a cell to produce a 'model cell'. In this model cell, the metabolic pathways are not considered to be individually regulated and the model is studied merely to give rough guidelines about general features of cell regulation. One such approach was adopted by Morowitz and his co-workers. They pictured part of the cell as a 'biochemical vat' with the following features:

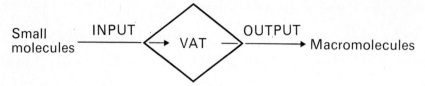

Figure 1 'Biochemical vat'.

1 The vat contains essentially a mixture of small molecules and class I enzymes.

2 'Input' into the vat is small molecules.

3 'Output' is macromolecules.

4 Class II enzymes help remove small molecules as 'output'.

5 Only a limited number of types of enzyme reaction (e.g. splitting enzymes) are considered to occur in the vat. No 'real' reactions are considered.

6 No control systems are considered.

A series of equations was constructed to predict the output from the vat under a variety of conditions. These equations were fed into a computer which then simulated (mimicked) the vat. The stability of the system under different conditions was examined.

Two interesting conclusions could be reached:
(a) Even without any control systems, the vat tends to be stable. (This is what Morowitz and his co-workers call 'passive stability'.)
(b) The more interconnections between the hypothetical pathways, the more stable the vat. Interconnections could be due to enzymes that link two or more pathways (e.g. hexokinase, Unit 4) or due to common intermediate compounds (e.g. ATP, NAD).

In terms of real cells, these features suggest that cells possess a stable base on which other 'active' control systems can be built. Further, the interconnections between pathways (which a glance at your 'Metabolic Pathways' chart in S100, Unit 15, will show are very numerous) help maintain this stability. This, however, is at a price, because more enzymes are needed to provide such an extensive network. The overall economy of the cell must therefore have evolved as a compromise between maximum efficiency (i.e. fewer enzymes, fewer reactions, and hence a more rapid transit from input to output) and maximum stability. However, as you will see from these Units, the degree of compromise between efficiency and stability is affected by other active control systems.

5.2.2 Techniques for studying the sequences of metabolic pathways

Study Comment

In this Section – as a prelude to discussing the regulation of metabolic pathways by controlling the activity of enzymes – we consider some experimental techniques employed in unravelling the sequences of reactions in metabolic pathways. You are not required to remember any of the specific examples that we give as illustrations, but should know the main principles of the techniques in order to achieve Objective 3.

There is a variety of techniques that can be employed in unravelling a metabolic pathway. Some of these, such as the use of radioactively labelled compounds as tracers (Unit 4, and its TV programme), you have already learnt about. Two techniques that have proved most successful in determining sequences of reactions, which we consider here, are isotope competition and the use of metabolic mutants. The latter technique has mainly been applied to bacteria and fungi. It is from *E. coli*, where both techniques have been used with great success, that we shall tend to choose our examples.

(a) *Isotope competition*

If a cell is supplied with a compound in which the atoms are radioactive and the cell can metabolize that compound, the products of this metabolism will be radioactively labelled too. This principle lies behind the use of isotopes as tracers of metabolic pathways. The technique can be further sophisticated by specifically labelling particular atoms in the compound supplied to the cell. In this case, the metabolic fate of various parts of the initial compound supplied can be followed. You saw for example how this is used to determine the relative contributions of glycolysis and the pentose phosphate pathway in glucose metabolism (Unit 4).

E. coli can synthesize a huge variety of organic compounds. The carbon atoms for these compounds can come from a number of possible sources. The one most commonly employed in growing *E. coli* in the laboratory is glucose. That is, glucose can serve as the sole source of carbon for *E. coli*. Therefore, if *E. coli* is supplied with glucose in which the carbon atoms are radioactive (^{14}C), all of its carbon-containing compounds will soon be radioactively labelled. It is difficult to trace the exact route of conversion of the carbon from glucose to all these molecules, as the process is rapid and several different metabolic routes occur simultaneously. However, in the 1950s, a group of biochemists working at the Carnegie Institute in Washington overcame this problem by utilizing a technique known as *isotope competition*.

Consider a hypothetical pathway in *E. coli*, whereby glucose is converted to a substance X, via four intermediate compounds A, B, C, D. Assume that the chemical identity of A, B, C and D is known, but the order in which they arise in the pathway is not known. Suppose that, in fact, the order is:

$$\text{glucose} \xrightarrow{1} A \xrightarrow{2} C \xrightarrow{3} D \xrightarrow{4} B \xrightarrow{5} X$$

The numbers, 1–5, refer to the steps in the pathway and to the enzymes catalysing them, i.e. enzyme 1 catalyses step 1, enzyme 2 catalyses step 2, etc. We shall adopt this convention throughout this Unit.

If ^{14}C-glucose (i.e. glucose molecules in which all the carbon atoms are equally radioactively labelled) is added to the cells, all the intermediates (A, B, C and D) and X will become radioactively labelled. Later, some non-radioactive substance C (prepared in the laboratory) is added to the cells.

QUESTION Which one of the following alternatives would you expect after the addition of C?
1 The specific radioactivities (Unit 2) of A, C and X would fall, those of D, B and glucose would remain unaltered (i.e. with respect to cells to which no C is added).
2 The specific radioactivities of C, A, B and X would fall, those of glucose and D would rise.
3 The specific radioactivities of C, B, X and D would fall, those of A and glucose would remain unaltered.
4 The specific radioactivities of all the compounds except glucose would fall.

ANSWER The correct alternative is (3). On addition of non-radioactive C, assuming it enters the cells, it would mix with and hence lower the specific radioactivity of the C. Since C is then converted to D, B and X, these too would be of lower specific radioactivity than in a control set of cells to which no C is added. However, as glucose and A occur before C in the pathway, they are unaffected by the reduction in the specific radioactivity of C, D, B and X. Thus, this experiment shows that C must occur after A but before D, B and X in the sequence. Similarly, if the experiment were repeated, adding non-radioactive D instead of C, this would reduce the specific radioactivity of D, B and X, but not of A and C; and so on. In this way, one can determine the entire sequence. There are, however, limitations to this technique, as you will see.

(b) *Metabolic mutants*

If compound X, which (as we mentioned above) is produced from glucose as follows:

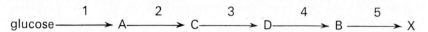

is a compound required by the cells for growth, then any mutant bacteria (i.e. those carrying a heritable alteration (S100, Unit 19)) that are unable to produce X will not be able to grow on glucose as the *sole* source of carbon. They will be able to grow on glucose plus X.

In the pathway from glucose to X, mutants could arise lacking any of the enzymes catalysing the five steps indicated. However, mutants lacking different enzymes will show different characteristics. For example, a mutant lacking enzyme 3 will, when given glucose, tend to accumulate substance C, whereas a mutant lacking enzyme 2 will accumulate A. Therefore, by isolating various mutants in a metabolic pathway and identifying the substances that they accumulate, one can identify some of the intermediates in the pathway. However, when one isolates a mutant for a particular pathway, one does not necessarily know which step of the pathway is blocked (i.e. which enzyme is lacking). Therefore, knowing all the intermediates that are accumulated in a particular mutant does not in itself give the order of the pathway. For example, consider a pathway where it is known that compound P is converted to compound Y, but the identity and order of the intermediates is unknown. Say six different types of mutant can be found, each mutant accumulating a different compound, which is then isolated and identified. We will arbitrarily label the mutants a–f. In Table 1 we give a list of the compounds that these mutants accumulate. If you try to determine the order of the pathway from P to Y from this data alone, you will find it impossible. Satisfy yourself that this is so.

Table 1

Mutant	Compound accumulated
a	R
b	Q
c	V
d	W
e	T
f	P

So all we can establish is that P → Y occurs via R, Q, V, W and T, in some unknown order. Now if Y is required for growth, the mutants will not grow on P alone, but will grow on P plus Y. Each mutant will also be able to grow on P plus those intermediates that occur after the blocked step (i.e. the step where the enzyme is lacking), since after the block the remaining enzymes can convert such intermediates to the required Y. In Table 2 we present data concerning the ability of mutants a–f to grow on P plus each of the intermediates in turn.

> **ITQ 1** From this and the data in Table 1, attempt to determine the order of the hypothetical pathway from P→Y. Assume that all the reactions from P to Y occur rapidly in the direction of Y and are not readily reversible.
>
> Do not spend more than 20 minutes attempting this question; check your answer against that given on p. 44.

Table 2

Mutant	Will grow on P plus	Will not grow on P plus
a	T or Q or W or V	R
b	V or W	R or T or Q
c	W	T or V or Q or R
d	none	Q or W or R or T or V
e	W or Q or V	T or R
f	R or Q or V or W or T	—

If you have attempted this question, or at least checked the answer, you should now have some idea of the rationale behind the technique. However, this technique has limitations, some of which it shares with the technique of isotope competition:

(i) Both techniques require that the intermediates are compounds which are stable enough to be either isolated or added to cells.

(ii) Both techniques require that the intermediates are taken up by the cells.

(iii) It is always possible that substances identified as intermediates by either technique are not really involved in the pathway under examination but 'feed into it'. For example:

$$\text{glucose} \xrightarrow{1} A \xrightarrow{2} C \underset{6}{\overset{3}{\rightleftarrows}} D \xrightarrow{4} B \xrightarrow{5} X$$
$$Y$$

Compound Y is not intermediate between glucose and X, but could be identified as such by using metabolic mutants or by isotope competition, because, if reaction 6 is readily reversible, a mutant lacking enzyme 4 would accumulate D, which would tend to be converted to Y (dynamic equilibrium; S100, Unit 12), which would then itself accumulate. (This back-reaction can, of course, also confuse data based on mutants in the true pathway, e.g. if, in the pathway $A \xrightarrow{1} B \xrightarrow{2} C \xrightarrow{3} D$, step 2 is freely reversible, a mutant blocked at step 3 will accumulate C and hence by back-reaction B.) Likewise non-radioactive Y would compete in isotope competition experiments following its conversion to D.

Objections (i)–(iii) can be overcome by the application of several experimental approaches, but the unravelling of any metabolic pathway is never a simple task (S100, radio programme of Unit 15).

5.2.3 Feedback inhibition in a linear pathway

Given sources of carbon, nitrogen, hydrogen, oxygen and sulphur, *E. coli* can synthesize, among a whole range of compounds, all the twenty amino acids that it needs to synthesize proteins. The pathways leading to these amino acids are highly complex, as yet another glance at your 'Metabolic Pathways' chart (S100, Unit 15) will doubtless convince you! Many higher organisms cannot synthesize all twenty amino acids, and so some—the so-called 'essential' amino acids*—have to be obtained from their food. The great metabolic versatility of *E. coli* has attracted many studies, and during some on the metabolic interconnections between various amino acids interesting control features have come to light.

For example, the amino acid isoleucine is synthesized from another amino acid, threonine; this in turn is synthesized from yet another amino acid, aspartic acid, which itself derives from glucose (or some other carbon sources) via the citric acid cycle (Unit 4). In *E. coli*, the conversion of threonine to isoleucine involves five enzymic steps which can be represented by:

$$\text{threonine} \rightarrow M \rightarrow N \rightarrow O \rightarrow P \rightarrow \text{isoleucine}$$

QUESTION If ^{14}C-threonine were added to some *E. coli* growing on glucose (non-radioactive) as a carbon source, and some time later protein was isolated from the cells, hydrolysed and the amino acids separated, which of the amino acids would you expect to be radioactively labelled?

ANSWER Threonine and isoleucine. Threonine, because some of the ^{14}C threonine added to the cells would be incorporated into protein. Some more of the ^{14}C-threonine would be converted to give ^{14}C-isoleucine which would then also be incorporated into protein.

While examining this pathway, both by isotope competition techniques and by using mutants blocked in the pathway, it became apparent that if a large excess of isoleucine (that is, much more isoleucine than the cells need to maintain their normal rate of protein synthesis) was added to the cells, then the cells ceased to convert any more threonine to isoleucine. So the cells only produce isoleucine when it is needed in larger amounts than already present. The end-product of

---

* *What is an 'essential' amino acid in one animal (i.e. it must be supplied in the food) is not necessarily an 'essential' one in another type of animal, since there may be minor variations between the synthetic capacities of different animals.*

the pathway, isoleucine, appears to inhibit the functioning of the pathway. In principle, isoleucine could inhibit the pathway, and thus its own synthesis, by inhibiting any one of the five enzymic steps.

> QUESTION Assuming that the only substances in the pathway that the cells need for growth are threonine and isoleucine, can you suggest from your knowledge of Unit 4 which step might be inhibited?

> ANSWER Step 1 – the conversion of threonine to M. This ensures that in the presence of excess isoleucine no wasteful conversion occurs of threonine to the (now) not needed intermediates M, N, O and P.

Indeed, in 1956, an American biochemist, Edwin Umbarger, made the important observation that Step 1, threonine→α-ketobutyric acid, (M in our scheme), is catalysed by an enzyme, threonine deaminase, which was inhibited *in vitro* by isoleucine. The inhibition was specific to isoleucine, other amino acids having much smaller or no effects. In a solution containing 10^{-2} M isoleucine, the inhibition was 100 per cent, at 10^{-4} M isoleucine, it was 52 per cent. Thus, from these *in vitro* observations, one could argue that with a high level of isoleucine inside the cell, threonine deaminase is inhibited, and less isoleucine is synthesized. As isoleucine is used up (say in synthesizing proteins), the inhibition is released, threonine deaminase operates and more isoleucine is synthesized. Thus the level of isoleucine regulates the synthesis of isoleucine—supply (by production) is geared to demand. This *feedback system* (S100, Unit 18, Unit 4 of this Course) ensures that the cell does not overproduce isoleucine, and thus waste threonine. Such a system is a *negative feedback system*, since the regulation involves a negative (inhibitory) effect of the end-product on an earlier step. There are many other examples known, in both bacteria and higher organisms, wherein the end-products of pathways inhibit the operation of the pathway in this way. You have already encountered several examples in Unit 4. The phenomenon is called *feedback inhibition* or *end-product inhibition*. The substance causing the inhibition (e.g. isoleucine) is called the *feedback* or *end-product inhibitor*. The enzyme inhibited is said to be *sensitive to feedback inhibition*. The analogous phenomenon is *activation* (Unit 4). In this, compounds that are either substrates of early steps in a pathway or related to substrates (starting metabolites) *activate* (that is, increase the catalytic activity of) an enzyme catalysing an early step in the pathway. An example that you encountered in Unit 4 was the activation of PFK by ADP and AMP. As in the case of PFK, a single enzyme can be subject to both activation and feedback inhibition. In such a case, these are opposing forces. The teleological account of activation and feedback inhibition is simple. Feedback inhibition ensures no overproduction when an end-product is already in adequate supply. Activation ensures a rapid utilization of a starting metabolite when it is in a large amount. In pathways such as glycolysis (Unit 4), where both types of control are operative, the controls must be finely adjusted to ensure a harmonious balance of activation versus inhibition, so as to allow the rate of operation of the whole metabolic pathway to be well tuned-in to the cell's needs.

feedback inhibition

activation

In these Units, we shall adopt the following conventions in diagrams representing feedback inhibition and activation.

(a) The feedback inhibitor or activator will be shown in red.

(b) The line joining the feedback inhibitor or activator to the enzyme reaction it affects will be shown in red also.

(c) Reactions are indicated by an arrowed line in black, as elsewhere in this Course.

(d) Where activation is occurring, the red line will end with an asterisk at the reaction affected (Unit 4). This indicates that the activation is of the enzyme that catalyses the reaction indicated by the black arrow.

(e) Similarly, inhibition is indicated by the red line ending in a broad red bar at the reaction affected (Unit 4). This indicates that inhibition is of the enzyme that catalyses the reaction indicated by the black arrow.

> ITQ 2 Using these conventions, sketch out the pathway from threonine to isoleucine, indicating that isoleucine inhibits the enzyme catalysing Step 1 – that from threonine to α-ketobutyric acid.

Now turn to p. 44 and check your answer before continuing with the Unit.

There are some other features of the regulation of the threonine-isoleucine pathway that have some points in common with many other systems showing end-product inhibition.

1 The end-product, isoleucine, inhibits Step 1. This prevents conversion of threonine to useless intermediates (e.g. M, N, etc.). As you saw in Unit 4, the control point of a pathway is often the first enzymic step unique to that pathway (e.g. PFK).

2 The end-product inhibitor, isoleucine, is structurally dissimilar from the substrate or products of the enzyme (threonine deaminase) it inhibits.

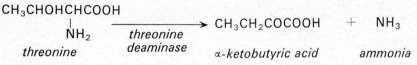

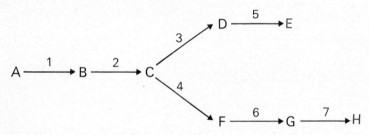

Enzymes like threonine deaminase, inhibited by end-products or activated by starting metabolites, are often termed *allosteric* (Greek: *allos*, other; *stereos*, solid). We shall discuss the chemical nature of such inhibition and activation in Section 5.3.

5.2.4 Feedback inhibition in branched pathways

The pathway from threonine to isoleucine is a linear one, there are no branches. Many metabolic pathways are branched. A simple branched pathway leading to end-products E and H is shown below:

$$A \xrightarrow{1} B \xrightarrow{2} C \begin{array}{c} \xrightarrow{3} D \xrightarrow{5} E \\ \xrightarrow{4} F \xrightarrow{6} G \xrightarrow{7} H \end{array}$$

QUESTION What would happen if one of the end-products, say E, when at a high level, inhibited Step 1?

ANSWER If the level of E built up, Step 1 would become inhibited, and this would lead to a reduced production of B and hence a reduced production of E. But it will also lead to a reduced production of H. This, of course, may prove 'undesirable' to the organism for, though it will be able to tolerate a reduced production of E, when E is at a high level, it may still require unabated production of H. The converse argument applies if Step 1 is inhibited by H rather than by E, when H is at a high level.

It is possible to sketch out on paper several theoretical ways of overcoming this problem. (You may like to try to do so for yourself. However, do not spend too much time attempting to construct control schemes—say 15 minutes or so—before reading on.) In practice, one finds that many of these ways are employed by living organisms. In different pathways and different organisms, some different mechanisms seem to operate. The overall result is, however, much the same – a more balanced production of all the end-products. In essence, the problem is to ensure that reduction of the production of one end-product does not reduce the production of other end-products of the same pathway. We now consider a few methods which overcome this problem.

(a) Enzyme multiplicity

In Unit 2 (Section 2.6), you considered the occurrence of isoenzymes—enzymes with similar catalytic activities but differing structures co-existing in the same organism or cell. In the example given, lactic dehydrogenase, the five types of isoenzyme all convert pyruvic acid to lactic acid, but are affected differently by the levels of the substrate (pyruvic acid) and the product (lactic acid) of the reaction. They were, therefore, subject to different controls. Isoenzymes, affected differently by different end-products of branched pathways, can form the basis of pathway regulation.

For example, consider the pathway shown below, where in a single cell two isoenzymes co-exist for catalysing Step 1 (A to B). One is feedback-inhibited by end-product E (isoenzyme 1e), the other by H (isoenzyme 1h).

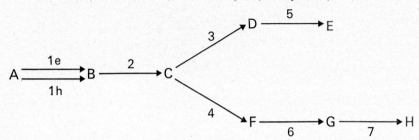

The consequences of such a scheme are very interesting. Suppose that a large amount of E gets into the cells. Isoenzyme 1e will be inhibited. Isoenzyme 1h will still operate, so B will be produced, but at a reduced rate (as only one isoenzyme is working). Reduced B will mean reduced C. So H will still be produced despite the high level of E. However, some of C will be converted to D (and D to E). Hence some of the output of isoenzyme 1h will lead to the production of E. Thus, the production of E will not be cut off completely, despite its high level, *and* the production of H (not at high level) may well be reduced. Likewise, an excess of H would not prevent some of the output of isoenzyme 1e being used in producing H. Such production would defeat the apparent purpose of feedback control—prevention of wasteful production. So other control points are needed as well as those operating at Step 1. If you now look at the branches for C to E (Steps 3 and 5) and from C to H (Steps 4, 6 and 7), you will notice that they form two linear pathways. So one obvious way of preventing a wasteful flow into one or other branch is to ensure that the first step in each branch (3 *or* 4) is also feedback-inhibited by the end-product of that branch.

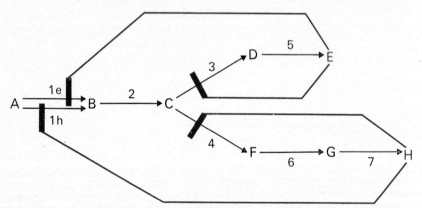

So if, for example, E is at a high level (i.e. in excess), it inhibits 1e. 1h still operates to produce B and hence C. But D is not produced from C, since Step 3 is also inhibited by excess E. So all the remaining C is diverted towards F, G and H. H production continues, though E production is reduced. The converse is true where H is in excess. Ideally, each isoenzyme is capable of converting enough A to B just to accommodate the necessary rate of production of the end-product to which it is sensitive. So, when one isoenzyme is inhibited, the normal output of the other end-product continues. You may well wonder why, if Steps 3 and 4 are inhibited by E and H respectively, the organism needs Step 1 to be inhibited also, because, if high E inhibits Step 3, then all of C is diverted to produce H. A similar argument applies for high H inhibiting Step 4. However, consider what would happen when both E and H are in excess. Steps 3 *and* 4 will be inhibited but if Step 1 is uninhibited by E or H, then all of A will be converted to C—a useless intermediate. Inhibition of Step 1 by E and H prevents this.

One example of a pathway in which the first step common to both branches (e.g. Step 1 above) is catalysed by isoenzymes is that in *E. coli* leading from aspartic acid to lysine and threonine. One isoenzyme is inhibited by lysine, and one by threonine. Later steps after the branch points are also feedback-inhibited. A simplified version of the pathway is shown below.

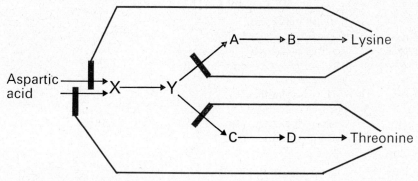

(b) Multiproduct inhibition

It is possible to achieve some balance of regulation in a branched pathway without isoenzymes at Step 1. One way is to have an enzyme catalysing Step 1 that is sensitive to more than one of the end-products, and ensure that no *one end-product alone*, even when in great excess, completely inhibits the enzyme. In some cases, high levels of both end-products (E and H) are needed to inhibit enzyme 1 at all (Fig. 2) *or*, though both E and H can inhibit enzyme 1 independently of the other, maximum inhibition is only achieved by both E and H being in excess (Fig. 3).

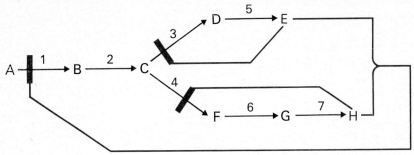

Figure 2

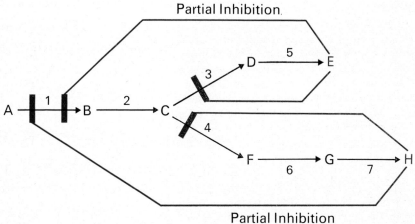

Figure 3

In both cases shown, Steps 3 and 4 are also inhibited by E or H respectively. Both of these systems ensure that production of both end-products will be severely inhibited only if *both* end-products are in excess.

For the schemes in Figures 2 and 3 and that discussed in (a), the enzymes catalysing Steps 3 and 4 must be sensitive to a similar level of E or H as that to which enzyme 1 (or 1e and 1h) is sensitive; otherwise some of C will be diverted to the wrong branch of the pathway. Some difference would be expected, however, in the sequence of events, depending on which scheme is operating. Suppose that a large amount of E becomes available to the cells, and that one of the schemes in Figure 2 and Figure 3 operates.

Try and work out for yourselves the sequence of events, which would occur if the scheme in Figure 2 operated. Repeat the exercise assuming the scheme in Figure 3 operated. Follow your reasoning as far as you can but do not spend more than about 20 minutes doing so. Then check your list of events against those in the table on p. 20.

Event based on Figure 2
1. Enzyme 3 inhibited.
   Normal production of C.
   All C diverted to path leading to H.
2. H builds up.
3. Enzyme 4 inhibited.
   Excess E and H together inhibit enzyme 1.
4. Pathway shuts down.

Event based on Figure 3
1. Enzyme 3 inhibited.
   Enzyme 1 partially inhibited.
   Reduced production of C.
   All remaining C diverted to path leading to H.
2. *If* H builds up, then . . .
   (a) Enzyme 4 inhibited. Further inhibition of enzyme 1 (by H).
   (b) Pathway shuts down.

Note particularly the 'if' in event 2 on the scheme in Figure 3. All subsequent events depend on it. H will only build up if the amount of C still synthesized is more than the normal amount available to this branch (F → G → H) of the pathway, when neither E nor H are in excess. In an 'ideal' (i.e. what *we* might *think* is ideal) system, E would only inhibit enzyme 1 to the extent where the loss of production of B and hence C was equal to the amount of C normally used to make E. Hence H production would continue normally, and on the scheme in Figure 3, events 2a and 2b would not occur.

(c) Sequential feedback

The synthesis of some amino acids in the bacterium *Bacillus subtillis* is regulated by a scheme like that outlined in Figure 4.

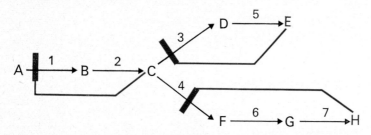

Figure 4

Neither end-product, E or H, inhibits Step 1. However, if one end-product, say H, is in excess, it inhibits the first step of its own branch of the pathway, Step 4. All of C will now be diverted to produce E. In consequence, E builds up and eventually inhibits Step 3. With 3 and 4 inhibited, C builds up. Enzyme 1 is sensitive to excess C and so Step 1 is now inhibited and the whole pathway shuts down. Here again, only when both E and H are at high levels does all synthesis cease. The high level of one end-product sequentially produces a high level of the other.

5.2.5 Evolution of feedback systems

We have now briefly reviewed some of the types of feedback systems whereby metabolic pathways can be regulated. The approach we have adopted is to treat each scheme from a theoretical standpoint—how could such a pathway be regulated, etc. . . .? We must, however, point out that though schemes of the type outlined above are found in a wide variety of living organisms, they were not necessarily discovered by looking for specific types of feedback system. Often the 'theory' of how it could be done has come after the discovery of how a particular regulatory system works in practice. We have presented this subject from a theoretical standpoint for convenience. It avoids having to give numerous examples, involving many complex compounds and metabolic pathways. It is after all the principles of how regulatory systems operate, rather than specific examples, that we wish you to understand.

It is now interesting to ask how such regulatory systems arose during evolution. One cannot, of course, give any definite answers. Nevertheless, by examining the various systems that operate in the organisms that now survive, one may gain some idea about the origins of these systems. The first thing to strike one is that the basic metabolic pathways, that is, those that are involved in producing compounds that are needed by all organisms, are very similar from one organism to another. Thus, in making a compound X from another compound A, most organisms go through a similar sequence of reactions, involving the same intermediate compounds. This suggests that these basic metabolic pathways were evolved early on, so that further evolution has not involved drastic changes in basic metabolism. Presumably any such changes would be detrimental and the organisms die out. The enzymes catalysing the individual steps carry out the same catalytic functions, but from organism to organism differ in their detailed structure.

It is on these differences in detail that different control mechanisms can operate. When one examines the regulation of these pathways, one finds that even the same pathway can be regulated in a variety of ways. Often, closely related organisms have different regulatory mechanisms for the same pathways. For example, in *E. coli* the regulation of the pathway from aspartic acid to lysine and threonine involves two isoenzymes at Step 1; the same pathway in another bacterium, *Rhodopseudomonas capsulatus*, involves only one enzyme at this step, as it is inhibited by a system like that shown in Figure 2. So two different regulation systems have evolved, each capable of regulating the same pathway. However, the selective pressure was probably the same – the pressure to maintain a balanced production of end-products, without wastage or underproduction.

We can perhaps try to reconstruct the evolution of a regulatory system from an unregulated pathway.

At some stage a mutation occurs that leads to an altered enzyme in the pathway becoming sensitive to a distant compound in the pathway, though its catalytic role is unaltered. If this new sensitivity (say feedback inhibition) enables the organism to regulate the pathway better, thus making the organism more efficient than its unaltered rivals, it will outgrow and out-reproduce them. Thus the 'new' altered organism will soon become the common type, i.e. it will have been selected. In this way, whole regulatory systems could evolve. However, since, as you have seen, there are several ways of achieving regulation, there is no reason why several regulatory systems should not evolve in parallel, and this is presumably what has happened.

Thus one of the fundamental steps in the evolution of feedback systems is probably the occurrence of enzymes sensitive to compounds other than their substrates and products, this sensitivity being reversible, i.e. dependent on the levels of the compound. We have so far in this Unit mainly discussed negative sensitivity (inhibition); however, positive sensitivity (activation) is also fairly common and can play a role in regulatory systems, as you saw in Unit 4 (e.g. PKK). In either case, it is interesting to inquire what are the features of the sentitive enzymes that render them controllable. It is this question that we tackle in the next Section.

SAQ for Sections 5.2.3–5.2.4

You should now attempt the following *SAQ*. (Answers on p. 46).

SAQ 1 (*Objective 4*) In a certain bacterium a metabolic pathway is regulated as follows:

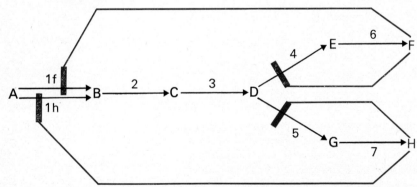

(a) If excess H is added to the cells, which of the eight compounds shown (A–H) will decrease in amount? (Assume 1h produces enough B to satisfy normal synthesis of H.)

(b) If a huge amount of D is added to the cells, what might be the subsequent sequence of effects on the synthesis of A–H in the cells?

(c) If a mutant form of the bacterium lacks isoenzyme 1f, what will occur to the levels of A–H when excess F is added to the cells?

5.3 Allosteric Inhibition

As we noted for isoleucine and threonine deaminase (Section 5.2.3), substances that are feedback inhibitors are not closely related chemically to the substrates or products of enzymes that they inhibit. These inhibitors work reversibly, so they are not just permanently damaging the active site, as do some non-specific inhibitors or quasi-substrates (Unit 2). It is, therefore, reasonable to postulate that these inhibitors do not bind at the active sites on the sensitive enzymes, but on other sites, so-called *allosteric* sites. If this is so, the fascinating question arises as to how the binding of a substance to one part of an enzyme can affect the functioning of the enzyme at another part (the active site). The answer is not yet known. However, there is good evidence that allosteric sites exist on those enzymes that are subject to feedback inhibition or activation, and there are some interesting hypotheses advanced to explain the behaviour of such enzymes.

allosteric sites

You should now read from the bottom of p. 258 ('Allosteric control of enzyme activity') to the end of Chapter 10 (p. 266) in *Cell Structure and Function*. Note the following:

1 You are not required to remember any details specific to the enzymes mentioned (e.g. ATCase). You should, however, as a result of studying the examples given, be able to evaluate evidence purported to support hypotheses concerning allosteric enzymes, as outlined in Objective 5.

2 You need not remember the exact details of the Koshland or Monod-Wyman-Changeux hypotheses. You should be able to point out differences between them (Objective 5).

3 Despite the legend to Figure 10–21, there are five steps, as we say in Section 5.2.3, between threonine and isoleucine.

4 Valine is an activator of threonine deaminase.

5 ATCase is not identical in detail to all enzymes subject to feedback inhibition. It has, however, points in common with other such enzymes.

6 The 'PHMB' at the top of Figure 10–23 should read PMB.

7 Do not worry about the S-shaped curve obtained when plotting the velocity of the enzyme reaction against substrate concentration (Fig. 10–26). The fact that it differs from the normal hyperbolic Michaelis-Menten curve (Unit 2; p. 239 of *Cell, Structure and Function*) is evidence for *cooperativity*. This cooperativity is one of the most interesting aspects of allosteric enzymes. In essence, it means that in enzymes having more than one active site (each site is assumed to be identical in structure), the binding of substrate to one site affects the subsequent binding of substrate to the other sites. Enzymes showing this property generate S-shaped curves (Fig. 10–26). Since desensitization of the enzyme to allosteric inhibitors or activators also destroys this cooperativity between *active sites*, it may be that inhibitors or activators operate on binding to the *allosteric sites* by affecting the degree of cooperativity between the *active sites*. This is elaborated on p. 262–5 of *Cell Structure and Function*.

Of course, both types of curve, S-shaped or hyperbolic, are still consistent with the formation of an enzyme-substrate complex (Unit 2).

5.4 Cell Compartmentation

So far, we have considered the effects of activators and inhibitors on enzyme activity. But the rate of an enzyme reaction also depends on the amounts of its substrates and cofactors that are present. As you already know from Units 1, 3 and 4, a cell is not merely a bag of substances. Enzymes in the same metabolic pathway are sometimes physically associated to form multienzyme complexes, such as the pyruvic dehydrogenase complex discussed in Unit 1. In such a complex, the product of one enzyme reaction is the substrate of another enzyme in the complex. The proximity of the enzymes to their substrates and cofactors ensures efficient operation of the pathway. It also removes the element of competition with other enzymes (not in the complex) in other pathways for substrates or cofactors produced in intermediate reactions in the complex, since many substrates or cofactors are common to several pathways (e.g. ATP, NAD). It is also likely that such complexes as a whole can be subject to regulatory influences.

These possibilities are very pertinent when considering higher organisms in which cell organelles, such as mitochondria and chloroplasts (Unit 3), occur. Such cells are divided into a series of compartments. Permeability barriers (Units 1, 3 and 4) exist between these compartments and the soluble fraction of the cell. So enzymes might be segregated from their substrates, cofactors, activators and inhibitors. This segregation cannot, of course, be absolute; the enzymes presumably encounter their substrates at some stage. However, often fairly complex systems are needed to overcome the segregation (e.g. the mitochondrial systems described in Unit 4) and such systems probably offer other control points for regulating metabolic pathways. Thus, by regulating the flow of metabolites in and out of the compartments by altering the permeability properties of the partitioning membranes or the rate of operation of the enzymes involved, the availability of substrates, cofactors, activators and inhibitors to various enzymes can be altered. So membranes are essentially impermeable barriers, and yet they do permit the flow of certain compounds across them. This flow is regulated—how, is not fully understood, but much attention has been directed towards the cell membrane, the membrane that surrounds the cell. It appears that this membrane contains a number of different proteins called *permeases*. Each type of permease is specific towards a particular compound, in much the same way that an enzyme is specific for its substrate or cofactors. Each permease can facilitate the passage of a particular compound across the membrane, thus allowing the compound to pass in or out of the cell. So, by regulating the activity or amounts of particular permeases in the membrane, the cell can control the rate of flow of specific compounds in and out of the cell.

permeability barriers

SAQs for Sections 5.3–5.4

You should now attempt the following questions:

> **SAQ 2** (*Objectives 1 and 5*)   Which of the following statements are true and which are false?
> (a) Allosteric inhibitors are always closely related chemically to substrates of the enzymes that they inhibit.
> (b) Enzymes showing cooperative effects have more than one active site.
> (c) The Monod-Wyman-Changeux model requires allosteric enzymes to be symmetrical.
> (d) All enzymic reactions occur in free solution.
>
> **SAQ 3** (*Objective 5*)   An enzyme catalyzing the first step in a six-step linear pathway (enzyme 1) is found to be inhibited by the end-product of the pathway, compound X. Which of the following would you expect to be true and which to be false?
> (a) Treatment of enzyme 1 with a substance destroying some amino-acid side-groups renders the enzyme catalytically unaltered but insensitive to X.
> (b) Enzyme 1 is irreversibly inhibited by X.

Now check your answers against those given on p. 46.

5.5 Summary of Unit 5

The following summary may serve to remind you of the major points made in this Unit.

1 Cells need to be regulated in order to grow and reproduce successfully.

2 Metabolic pathways can be regulated by controlling the rates of individual enzyme-catalysed reactions in the pathways.

3 The rate of any enzyme-catalysed reaction can be altered either by (i) affecting the catalytic activity of a fixed quantity of the enzyme or by (ii) changing the amount of enzyme present.

4 Metabolic pathways can be unravelled by many techniques including isotope competition and the use of metabolic mutants.

5 Feedback inhibition and activation are common ways of controlling both linear and branched pathways.

6 The control of branched pathways can vary in several ways.

7 Sensitive enzymes have allosteric sites for inhibitors or activators.

8 Cell compartmentation plays a role in the regulation of metabolism.

UNIT 6

6.1 Introduction

In Unit 5, we discussed how enzymes can alter their rate of production of various compounds in response to the level of these or related compounds present in the cell. The mechanisms of feedback inhibition and activation are such that the alterations in production can occur rapidly and be quickly reversed. For example, if in *E. coli* the level of isoleucine is in excess, threonine deaminase is inhibited and further production of isoleucine ceases quickly. When the level of isoleucine drops (owing to its being used up in protein synthesis) to below the level required for optimum protein synthesis, this level is insufficient to inhibit threonine deaminase and isoleucine production rapidly resumes. It would appear that feedback inhibition and activation have evolved because they possess the advantage of dealing effectively with short-term fluctuations in metabolite levels. But what of long-term changes? Will, for example, a bacterial cell growing in a medium containing a high level of an end-product still continue to make the enzymes of that metabolic pathway so that on division each daughter cell has the same amount of those enzymes as the parent cell? In other words, will a cell always have the same amount of an enzyme even when that enzyme is not being used? Consider the following example.

In *E. coli* the amino acid tryptophan is made from glucose as a carbon source via a long branched pathway. The last (unbranched) portion of the pathway involves the conversion of anthranilic acid to tryptophan via four enzymic steps:

$$\text{anthranilic acid} \xrightarrow{1} B \xrightarrow{2} C \xrightarrow{3} D \xrightarrow{4} \text{tryptophan}$$

As you might anticipate by now, the first step, that catalysed by the enzyme anthranilic synthetase, is found to be inhibited *in vitro* by tryptophan. So here again, feedback inhibition *could* explain the cessation of tryptophan synthesis that is observed on adding excess tryptophan to *E. coli*. But what of *E. coli* growing and dividing in the continuous presence of excess tryptophan? Examine the following experiment.

Two identical batches of *E. coli* (A and B) are grown on glucose as sole carbon source. Then to one batch only (Batch A), an excess of tryptophan is added. Both batches are allowed to grow and divide further. At various times after addition of the tryptophan to Batch A, samples of cells are taken from each batch, broken, dialysed to remove any tryptophan and assayed for anthranilic synthetase. An equal number of cells is taken in each sample, yet the extract from Batch A has a much lower specific activity of anthranilic synthetase than that from Batch B. This lowered specific activity cannot be due to feedback inhibition in the assay, since tryptophan has been removed by dialysis. It must, therefore, be *because the cells in Batch A contain less enzyme per cell than those in Batch B*. It appears that growth in the presence of excess tryptophan has led to a reduction in the amount of anthranilic synthetase in the cells. Further evidence that this effect is different from feedback inhibition (where the amount of the enzyme is unaltered) comes from the observation that the enzymes catalysing Steps 2, 3 and 4 are also found in lesser amounts in the cells of Batch A. (Feedback inhibition by tryptophan affects only anthranilic synthetase.) This reduction in the amount of the enzymes of a metabolic pathway in response to an end-product of the pathway is a fairly widespread phenomenon in bacteria and other microorganisms. It is called *end-product repression* or *enzyme repression*. Many pathways that show end-product repression also show feedback inhibition and vice versa. Like feedback inhibition, end-product repression is also reversible. So, if some of the cells of Batch A are washed free of tryptophan and put in a medium with glucose and without tryptophan, a rapid increase in their content of anthranilic synthetase and enzymes 2–4 occurs. If the cells are then split into two batches (A1 and A2) and tryptophan is added to one only (A1), then further increase in the levels of these enzymes soon ceases (Fig. 1) due to renewed repression.

enzyme repression

Though they operate by different mechanisms, feedback inhibition and end-product repression act in the same direction. They both reduce production by metabolic pathways. Many pathways are controlled by both mechanisms. It is

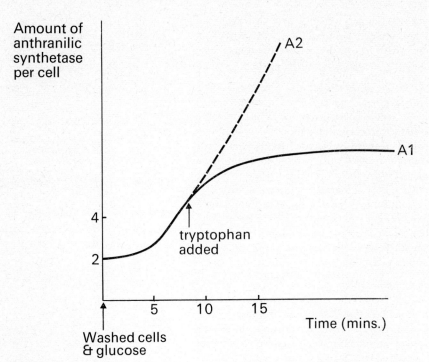

Figure 1

interesting to speculate on the relative contributions of each mechanism in such cases. Because end-product repression depends on a decrease per cell in the amounts of enzymes, it has its effect only slowly. Thus, in bacteria, the level of enzyme will drop only at cell division (i.e. if no new enzyme is made prior to division, then each daughter cell has 50 per cent of the level in the parent cell), say by 50 per cent every 30 minutes. Feedback inhibition, on the other hand, operates very rapidly to shut off the production of end-product. It, however, does not lead to an economy on enzyme levels. Therefore, a combination of both mechanisms gives much tighter control of the pathway: end-product production is rapidly shut off and enzyme levels are slowly reduced. Reversal of feedback inhibition (when excess end-product is used up) is also more rapid than complete reversal of end-product repression. So feedback inhibition is a rapid mechanism; repression is slow but has the advantage of economizing on enzymes.

If end-product repression can be regarded as a formal analogue of feedback inhibition, is there an analogue of activation (Unit 4)? In fact, such a phenomenon does exist and is called *enzyme induction*. It occurs in some catabolic pathways, that is, pathways wherein complex organic molecules are broken down to yield simpler ones (Unit 4). It will not surprise you that we return once again to *E. coli*, where the classical example of this phenomenon has been described.

enzyme induction

Though *E. coli* can use one of a variety of organic compounds as its sole source of carbon, the best source is glucose (*E. coli* grows fastest when using glucose as carbon source). The glucose is converted via glycolysis and the citric acid cycle (Unit 4) to provide the carbon for all the organic molecules in the cell. All of these conversions require specific enzymes (Unit 2) which are themselves synthesized by *E. coli* using glucose carbon. Lactose is a disaccharide containing one molecule of galactose linked by a β-galactoside link to one molecule of glucose. To metabolize lactose, *E. coli* must first cleave the β-galactoside link to release the glucose and galactose, which can then be fed into the glycolysis pathway. The cleavage of the β-galactoside link is achieved by means of an enzyme, β-galactosidase (Fig. 2).

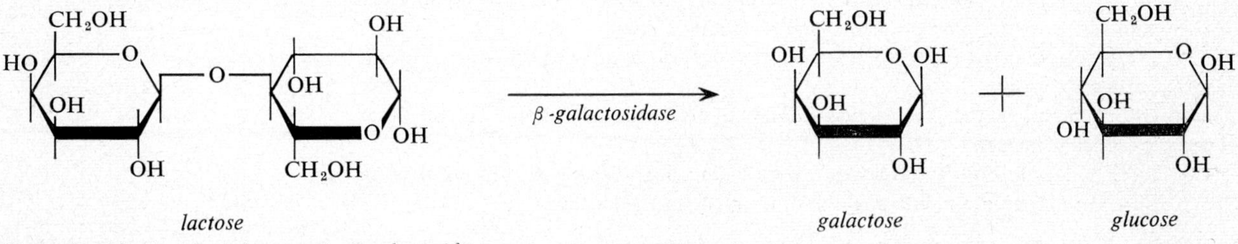

Figure 2 *The cleavage of lactose by β-galactosidase.*

E. coli grown on glucose as sole carbon source, where lactose cleavage is not needed, has very little β-galactodidase present. However, if *E. coli* growing on glucose is transferred to a medium where lactose is sole carbon source, then the amount of β-galactosidase in the cells increases rapidly (Fig. 3).

This increase in the *amount* of an enzyme (β-galactosidase) in response to the presence of its substrate (lactose) is called *induction*. It is due to an increase in the amount and not in the catalytic activity of the enzyme. Thus induction resembles end-product repression. Induction occurs in response to a high level of substrate and leads to an increased quantity of enzyme. Repression occurs in response to a high level of end-product and results in reduced quantities of enzyme. Release of repression when end-product is removed (Fig. 1) directly resembles induction. An enzyme that increases during induction is said to be *induced*. The substance causing the induction (substrate or related compound) is called an *inducer*. Enzymes that are subject to induction are said to be *inducible*. (Those subject to end-product repression are *repressible*.) Induction and repression do not involve any permanent change in the bacteria, as they are readily reversible (Fig. 1), e.g. remove the inducer and the enzyme ceases to increase.

Induction is a common phenomenon, and though, as you will see, its detailed basis has only been unravelled in recent years, the basic observations are very old. It is, for example, well-known to brewers that yeasts need 'training' – yeasts grown on one sugar need a period of adjustment before growing on another type of sugar. This 'training' period is presumably when enzymes responsible for converting the 'new' sugar are induced. This 'training' was examined scientifically as long ago as 1900.

Though we have limited our discussion of induction and repression to microorganisms, these phenomena are also observed in all types of higher organisms. Indeed the induction of β-galactosidase in the pancreas of the dog on feeding lactose to the animals was investigated around the same time as was induction in yeast (1900). Other changes in amounts of particular enzymes present in various tissues can be effected by hormones or foodstuffs. We shall discuss these phenomena in more detail later in this Unit, but in the main we shall adhere to our policy of choosing our examples from bacteria where more is understood, and then examine higher organisms by comparison. How then does enzyme induction come about?

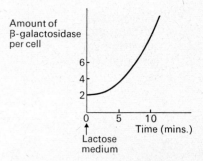

Figure 3 Induction of β-galactosidase

6.2 Synthesis and Degradation of Proteins

Enzymes are proteins. Proteins, like other macromolecules in living organisms, are synthesized by those organisms. This synthesis, as you know, is not a 'one-off' job. Proteins are continually synthesized, degraded, and re-synthesized (S100, Unit 15). In growing cells, more protein is synthesized than degraded. In non-growing cells there is a balance (a steady-state) wherein the rate of protein synthesis equals the rate of protein degradation.

As you will recall from S100, Unit 17, protein synthesis involves three steps: the transcription of DNA to give specific mRNAs; the translation of those mRNAs to give specific polypeptides; and the folding of one or more polypeptides to give the active three-dimensional structure of the protein (Units 1 and 2 of this Course). Degradation, as applied to enzymes, means a complete loss of enzyme activity due to a denaturation of the active shape of the protein (S100, Unit 14) or the breaking of peptide bonds so as to reduce the protein to the component amino acids.

In principle, any change observed in the amount of a particular enzyme in a cell could be due to a change in the steady-state as a result of one of three causes:
(a) a change in the rate of synthesis of that enzyme relative to the other proteins in the cell;
(b) a change in the rate of degradation of that enzyme relative to the other proteins in the cell;
(c) changes in both the rates of synthesis and degradation of that enzyme, of differing magnitude, relative to the other proteins in the cell.

Thus, induction of an enzyme could be due to either an increased rate of synthesis of the enzyme *or* a decreased rate of degradation *or* a combination of the two. Similarly, enzyme repression could involve a decrease in the rate of synthesis of the enzyme *or* an increase in the rate of degradation *or* a combination of the two. There is no guarantee that one unique mechanism exists for all enzyme induction and repression. Different enzymes in the same organism, or the same type of enzyme in different organisms, may be induced or repressed by different mechanisms. Studies on bacteria have revealed that degradation of protein, though it occurs sometimes, tends to be non-selective and therefore is unlikely to account for the specific changes seen on induction or repression in bacteria. It is likely, therefore, that in bacteria these phenomena mainly involve a control of protein synthesis. However, in higher organisms, the situation is more complex. Induction and repression occur more slowly and protein degradation is more widespread and rapid. So in these organisms, all three mechanisms (a)–(c) could contribute significantly towards induction and repression. It is interesting to speculate on the evolutionary significance of this difference in protein degradation between bacteria and higher organisms. In bacteria, if enzyme repression operates by inhibiting further synthesis of that enzyme, then the level of enzyme per cell will decrease fairly rapidly as the bacteria divide. Thus, at each cell division, the daughter cells will contain half the quantity of the enzyme that the parent cell contained. So in *E. coli*, dividing say once every 20 minutes, a rapid exponential fall in the level of the repressed enzyme will occur (like an isotope decaying (S100, Unit 2)): after 20 minutes, the level per cell will be 1/2, at 40 minutes 1/4, at 1 hour 1/8, at 2 hours 1/64, and so on. However, in higher organisms, some cells divide only rarely—once every few days, or weeks, or even years. If such cells could reduce the level of an enzyme only by inhibiting further synthesis and relying on dilution by cell division, this would be a very slow way of achieving a reduction in enzyme level. Hence the possible advantage of a mechanism for degrading proteins specifically.

As you will no doubt by now guess, more is understood about the detailed mechanisms of induction and repression in bacteria than in higher organisms. Since we have already argued that these phenomena in bacteria are mainly achieved by regulating protein synthesis, we shall first consider some of the possible ways of regulating protein synthesis, their relevance to induction and repression in bacteria and then their relevance to higher organisms.

SAQs for Sections 6.1 and 6.2

You should now attempt the following questions:

> **SAQ 4** (*Objectives 1 and 5*) Some *Bacillus subtilis* (a bacterium) growing on glucose as sole carbon source were transferred to a medium where the amino acid histidine was sole carbon source. After a few minutes an increased amount of the enzyme histidase (an enzyme capable of degrading histidine) could be detected in the cells.
>
> (a) The increased level of histidase is due to:
>   (i) Enzyme repression by histidine.
>   (ii) Feedback inhibition by histidine.
>   (iii) Enzyme induction by histidine.
>   (iv) Competitive inhibition by histidine.
>
> Choose the best alternative.
>
> (b) Histidine, with respect to histidase is:
>   (i) An inducer.
>   (ii) A feedback inhibitor.
>   (iii) A quasi-substrate.
>   (iv) A competitive inhibitor.
>
> Choose the best alternative.
>
> **SAQ 5** (*Objectives 3 and 5*) (a) *E. coli* growing on radioactive glucose produces radioactive arginine (an amino acid). A few minutes after addition of non-radioactive arginine to the cells, no radioactive arginine can be detected. Based on this evidence alone, this could be due to:
>   (i) Feedback inhibition of the first enzyme of the arginine biosynthetic pathway by arginine.
>   (ii) Repression of the synthesis of the enzymes of the biosynthetic pathway by arginine.
>   (iii) A combination of (i) and (ii).
>   (iv) None of the above.
>
> Choose the best alternative.

(b) Consider the evidence in (a) plus the following:

On removing the non-radioactive arginine from the batch of cells, radioactive arginine can be detected in the cells within a few seconds. If puromycin, a potent inhibitor of all protein synthesis, is added to the cells at the same time that the non-radioactive arginine is removed, rapid resumption of synthesis of radioactive arginine still occurs. Which of the following alternatives now explains the resumption best?

(i) A release of enzyme repression.
(ii) A release of feedback inhibition.
(iii) Increased overall protein synthesis.

Choose the best alternative.

Now check your answers against those given on p. 46.

6.3 The Mechanism of Protein Synthesis

Study Comment

This Section depends on reading the appropriate parts of *Cell Structure and Function*. Much of the material covered in this Section has been covered in less detail in S100, Unit 17. If you are very familiar with Unit 17, and *if* you are short of time, you might be able to skim through this Section quickly. You should in any case try the SAQs at the end of the Section. Once again you are not required to remember masses of details, but you should know the following:
1 where RNA is made in the cell;
2 where protein is made in the cell;
3 the basic structure of ribosomes;
4 the components necessary for synthesizing protein from mRNA in a cell-free system.

In the previous Section, we argued that the phenomena of induction and repression in bacteria were more likely to involve control of protein synthesis than control of protein degradation. Our definition of protein synthesis was a broad one, encompassing all the events from the transcription of DNA to give mRNA to the spontaneous folding of polypeptides to give the specific shape of active protein. Before we can discuss control of protein synthesis in any detail, it is therefore necessary to review what is known about the basic mechanism of protein synthesis. This mechanism is very complex but nevertheless appears to be fundamentally the same in bacteria and higher organisms. Transcription (the copying of DNA sequences into RNA) has already been adequately dealt with in Unit 17, but translation was only mentioned briefly. You should therefore now read pp. 371–95 in *Cell Structure and Function*, noting the following:

1 You are not required to remember all the details given in this section of the book, except where we indicate below.

2 Table 15–2 (p. 378) gives the components necessary for the synthesis of protein with ribosomes (plus mRNA) from a rat liver cell-free system. In general, cell-free systems require:

Ribosomes (plus mRNA).
Supernatant (or pH5) fraction
ATP
GTP
Mg^{2+} and K^+ ions.
20 amino acids (if one or more amino acids used are radioactive, this enables the experimenter to detect the small amount of protein made).

The pH5 fraction is just the cell supernatant treated to reduce its pH to pH5 (to remove some unnecessary protein). Supernatant or pH5 fraction contain the necessary tRNAs, activating enzymes and transfer enzymes.

You should be able to recall the components listed above.

3 All you need remember about the structure of ribosomes is that those from bacteria are smaller than those from higher organisms, that all ribosomes consist of two subunits (one large and one small), and that each subunit contains RNA (ribosomal RNA) and protein.

4 You should understand the 'cycle' by which amino acids are joined together on the ribosome, as given in Figure 15–4 and S100, Unit 17.

5 You need not know the detailed steps of attachment of amino acids to tRNAs as shown in Figure 15–6. Just remember that the attachment of a particular amino acid to its specific tRNA requires ATP and an activating enzyme specific for that amino acid and for the corresponding tRNA.

6 Read the section on pp. 386–8 ('Specific Protein Synthesis'). It may appear a bit compact but we take up some of the points in the remainder of this Unit.

7 The section on pp. 388–95 ('Nucleic Acid "Code"') plus Table 15–8 on p. 396 should be read, but do *not* try to remember any of the codons. Note that the code given in Table 15–8 is almost certainly correct, as it has been confirmed by more recent studies.

8 You should understand the role of N-formylmethionyl tRNA as an initiator of new polypeptide chains. This special tRNA ensures that the mRNA is read from the correct starting point (AUG or GUG). This is particularly important where the mRNA contains information for making several different polypeptide chains along its length, and hence has several starting points. A new polypeptide chain can be started at each initiator codon (AUG or GUG). You need *not* remember the initiator codons, or the terminator ones (UAG, UAA and UGA).

When you have completed pp. 371–95 in *Cell Structure and Function*, look at Figure 4, which serves as a summary of the complete process of protein synthesis from DNA to active protein.

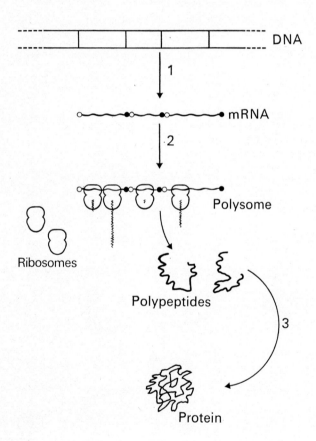

Figure 4 From DNA to protein: the synthesis of specific proteins.

Step 1 Transcription
Here we show three adjacent genes (indicated by cross-bars) on the DNA* which have been transcribed by DNA-dependent RNA polymerase to give a single mRNA molecule. This mRNA therefore contains three initiator and three terminator codons. These are indicated for the sake of clarity by unfilled circles (initiator) and filled circles (terminator). In reality, neither the ends of the genes in the DNA (cross-bars) nor the ends of the transcripts in the mRNA (circles) (by transcript in mRNA we mean the region corresponding to one gene in the DNA and hence containing the information for one whole polypeptide chain) are physically distinguishable from the rest of the molecule, except by a particular codon or sequence of codons.)

Step 2 Translation
2a Amino acid activation needs activating enzymes, ATP, tRNAs and 20 different amino acids.
2B Initiation needs N-formylmethionyl tRNA, ribosomes, GTP, initiator proteins.
2c Peptide bond formation needs mRNA, GTP, transfer enzymes, amino acyl tRNAs, ribosomes.
2d Termination – release of polypeptide chain – may require specific terminator enzymes that recognize the terminator codons in the mRNA and assist the release of the completed polypeptide.

Step 3 Folding
Folding of polypeptide chain(s) to give the active shape of the protein. (This probably occurs spontaneously and is dependent on the sequence of amino acids in the polypeptides.)

* At any particular region in the DNA probably only one of the two strands is transcribed to give mRNA.

SAQs for Section 6.3

You should now attempt the following questions:

SAQ 6 (*Objective 7*) Which of the following components are necessary, among others, for synthesis of proteins in a cell-free system?
(a) GTP; (b) membranes; (c) activating enzymes; (d) ATP; (e) glucose; (f) ribosomes; (g) amino acids; (h) lysosomes; (i) lactose.

SAQ 7 (*Objective 6*) Which of each of the following statements are true and which are false?

(a) Microsomes are ribosomes obtained from the cell nucleus.

(b) tRNA contains all the information for making proteins.

(c) N-formylmethionyl tRNA acts as an initiator in translation.

(d) Ribosomes contain only RNA.

(e) Each activating enzyme is specific for a particular amino acid and a corresponding tRNA.

(f) The genetic code is different from organism to organism.

SAQ 8 (*Objective 6*) When polysomes are treated with a certain enzyme they are disrupted to give single ribosomes. Is the enzyme likely to be (a) a ribonuclease (i.e. an enzyme that degrades RNA to nucleotide bases); or (b) a lipase (i.e. an enzyme that digests fats); or (c) a peptidase (i.e. an enzyme that splits peptide bonds)?

Now check your answers against those given on p. 46.

6.4 The Mechanism of Induction and Repression in Bacteria

Since the phenomenon of induction was first clearly described in the 1930s, there have been many attempts to explain it. One of the earliest hypotheses made use of the observation that the inducer of an enzyme was often a substrate of the enzyme. It assumed that an equilibrium existed in the cell between the active enzyme and an inactive pro-enzyme, that is a fully synthesized polypeptide chain or chains not folded into the active enzyme shape:

$$\text{pro-enzyme} \rightleftharpoons \text{enzyme}$$

In the absence of substrate (inducer), the equilibrium was far to the left-hand side of the equation. Substrate (inducer) was thought to form a complex with the enzyme and shift the equilibrium over to the right, and hence induction:

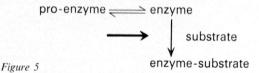

Figure 5

ITQ 3 Look back at Figure 4. Try to predict which components of the scheme depicted in Figure 4 would be necessary and which unnecessary, if induction was mediated by the mechanism shown in Figure 5. Try to design an experiment to test this mechanism.

To test your ideas, indicate which of the statements listed below would be true and which would be false, *if* the mechanism shown in Figure 5 were true:

(a) If radioactive amino acids were given to the cell during induction, no radioactivity would be detected in the induced enzyme.

(b) Inhibitors of peptide bond formation (i.e. the joining together of amino acids on ribosomes) would inhibit induction.

Now check your answers against those given on p. 44.

From this question you can see that the mechanism shown in Figure 5 could be tested in one of two ways. In fact, when many systems showing induction (e.g. β-galactosidase) have been tested in these ways, they have been shown to involve peptide bond formation—that is, the induction requires synthesis of new polypeptide chains from amino acids. Thus induction is inhibited by inhibitors of peptide bond formation and radioactivity *would* be detected in the induced enzyme, if radioactive amino acids were given to a cell during induction. Similarly in bacterial systems showing repression, it can be demonstrated that the increase in enzyme that occurs following removal of end-product (Fig. 1, p. 25) also requires synthesis of new polypeptide chains. It is now generally assumed that the phenomena of induction and repression in bacteria always involve the synthesis of new polypeptide chains—that is, induction involves an increase in the synthesis of the polypeptide chain(s) of the enzyme induced, and repression a decrease. (Strictly speaking one should reserve judgement and say that in all cases so far investigated the synthesis of new polypeptide chains is involved.) This probably reflects the fact that most polypeptides seem to fold up to give active proteins automatically and so a slow pro-enzyme ⇌ enzyme equilibrium is probably a rare event.

We can now rephrase the problem; how, during induction, is the rate of synthesis of a particular polypeptide chain controlled? If you look at Figure 4, you will see that there are many possible variables (and hence controllable components) in the system. However, many of these—such as ATP, GTP, tRNAs, amino acids—are common to the synthesis of all polypeptides and hence if altered in amount will cause changes in the rate of synthesis of all polypeptides (and hence all proteins). Any one induction or repression is a process specific to one or a few proteins. Therefore, only components specific to those proteins can be involved in these phenomena. An examination of Figure 4 will reveal only two such components: mRNAs and the genes on the DNA from which they were transcribed. Thus control of transcription must be one possible way of regulating the synthesis of specific polypeptides. Other controls could be exercised by altering the rate of the translation of mRNA on the ribosomes to give polypeptides. There is good evidence that in bacteria all mRNAs are very unstable, having half-lives of about 1 to 3 minutes. This means that, in principle, the switching on or off of transcription of certain genes would be a rapid way of switching on or off the synthesis of particular polypeptides. So, if transcription of a certain gene were switched off, no more mRNA would be made from that gene while switched off and consequently the existing mRNA from that gene (made before it was switched off) would soon be destroyed and hence no more polypeptides could be translated from that mRNA. This *could* explain the rapid halt in enzyme synthesis seen on adding end-product in repression (Fig. 1, p. 25), or on removal of inducer in induction.

Thus, one is faced with two alternative hypotheses to explain induction and repression:
1 Control is effected by controlling the rate of translation of particular mRNAs.
2 Control is effected by controlling the rate of production of particular mRNAs (transcription).

It is now known in several cases which of these two hypotheses is correct. However, most of what is now known about induction and repression has come from a detailed study of just one system—the induction of β-galactosidase in *E. coli* (Section 6.1). For over 30 years, the unravelling of this system has occupied many scientists, but its importance as a model for other systems far outweighs the apparent overworking of this one system. Upwards of a thousand scientific papers have appeared on the system. We obviously cannot discuss all the data in detail. What we shall do is briefly discuss the main observations, show how they point towards certain mechanisms, and then indicate which one of the two hypotheses is supported by the available evidence.

6.4.1 The lactose system

The basic observations are as follows:

1 On addition of lactose, or some non-metabolized analogues of lactose, to *E. coli*, an increase is observed in the cellular level of the enzyme β-galactosidase. The increase starts within a few minutes of adding the inducer (Fig. 3). The induction can result in up to a 1000-fold increase in the level of β-galactosidase.

2 If radioactive amino acids are added along with the inducer, the β-galactosidase subsequently formed is found to be radioactive. This rules out the possibility that the induction merely involves refolding of a protein, i.e. a pro-enzyme (Fig. 5).

3 Some analogues of lactose, though inducers, were found not to be substrates of β-galactosidase. This also rules out the scheme shown in Figure 5.

4 At the same time as inducing β-galactosidase, lactose also induces two related enzymes (galactoside permease and transacetylase); both can use lactose

as a substrate. Thus induction, like repression, leads to altered levels of several enzymes related through their metabolic functions.

5   The genes for all three enzymes were found by genetic techniques to be consecutive on the DNA. All three genes are transcribed together to give mRNA molecules containing the information from the three genes (Fig. 6).

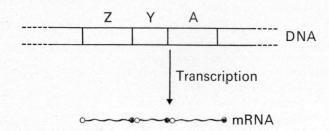

Figure 6   Genes for the three enzymes transcribed to give one mRNA: Z is the gene for β-galactosidase; Y is the gene for galactoside permease; A is the gene for transacetylase. The same convention is used as in Figure 4 for the ends of the genes in the DNA and the ends of the transcripts in the mRNA.

6   The mRNA corresponding to the three genes was found to have a half-life of about $1\frac{1}{2}$ minutes. This means that a rapid switch-off of synthesis of the three enzymes *could* occur following a switch-off of transcription.

7   Mutants of *E. coli* were isolated that make a high level of all three enzymes even in the absence of inducer. These mutants are called *constitutive*, as they make the enzyme as part of their constitution. Normal *E. coli*, which only makes the enzymes in the presence of inducer, is said to be *inducible*, and the ability to be induced is called *inducibility*. The enzymes made by the constitutive mutants were found to be identical in structure to those made by inducible cells.

constitutive mutants

8   The mutation leading to constitutive synthesis of the enzymes was shown to be located in a gene close to, but separate from, the genes for the three enzymes. It was called the *i* gene (for inducibility). This *i* gene must be a *regulator gene* as it is concerned with the regulation of the synthesis of the enzymes, not with their structure (see point 7). Accordingly, the genes for the three enzymes are called *structural genes* (Z, Y and A in Fig. 6).

regulator gene

structural genes

Having gathered this kind of data, and by using further genetic experiments, two French scientists, working at L'Institut Pasteur in Paris, François Jacob and Jacques Monod were able in 1961 to propose a hypothesis to explain enzyme induction. In Figure 7, we present a summary, in the form of a diagram, of their now famous hypothesis as applied to the lactose system, with one or two features added from more recent experiments. In order to understand this diagram fully, you should read carefully the description given below.

Jacob-Monod hypothesis

The hypothesis briefly stated is that the *i* gene contains the information for a protein called the *repressor*. This repressor, in the absence of inducer, binds to the operator, a small region of DNA adjacent to the Z gene and between it and the promoter. This repressor thus prevents RNA polymerase proceeding from the promoter on to transcribe the DNA of the Z, Y and A genes. No mRNA is made from these genes and hence none of the three enzymes (7A). Inducer, when present, can bind to repressor and prevent it binding to the operator region (7B). RNA polymerase molecules can then proceed to transcribe Z, Y and A many times. mRNA is thus made and translated to give the three enzymes (7C).

repressor

Jacob and Monod called a system where several enzymes are regulated together and the structural genes for these enzymes are consecutive on the DNA an *operon*. Thus, they considered that for each operon there is a regulator gene (e.g. the *i* gene for the *lactose operon*) which produces a repressor. This repressor is specific for the operator region of that operon and the inducer. Thus control is *negative*, i.e. regulation is achieved by preventing transcription, *unless* inducer is present.

operon

operator region

You should now examine Figure 7 *very carefully* and satisfy yourselves that it is compatible with all the data for the lactose system given in Section 6.1 and in points 1–8 above.

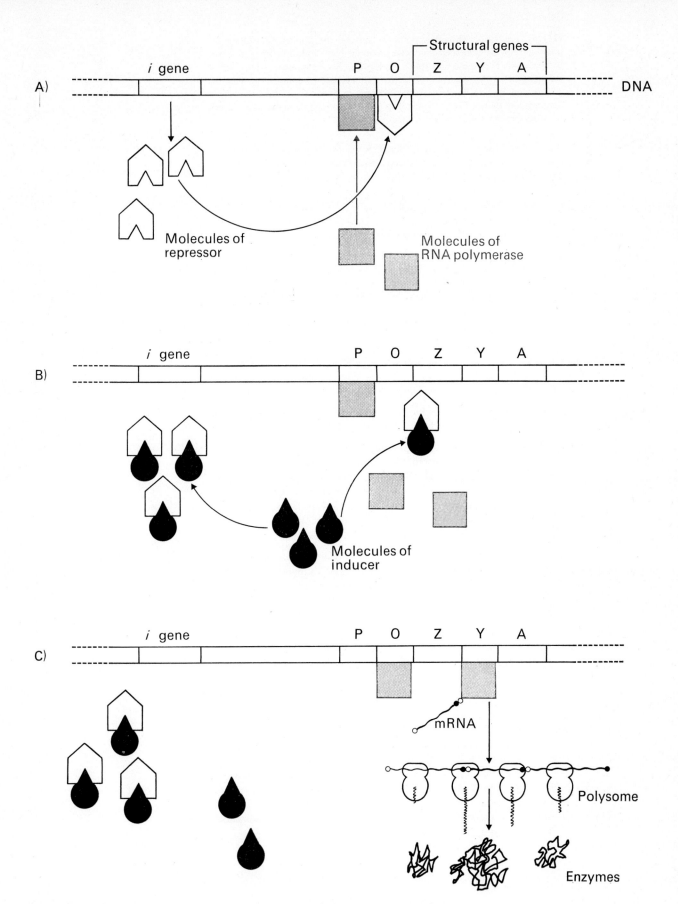

Figure 7 The Jacob-Monod hypothesis. Z, Y, A are as in Figure 6. P is the promoter, the region on the DNA where the enzyme responsible for copying DNA sequence into RNA (transcription), RNA polymerase, attaches. O is the operator region.

ITQ 4 The Jacob-Monod hypothesis can also be applied to explain repression. Having examined the lactose system, you should try for yourselves to apply this hypothesis to explain repression. Try mapping out in terms of the hypothesis, the sequence of events that occur on adding tryptophan to *E. coli* and lead to the repression discussed in Section 6.1. Constructing a diagram like that in Figure 7 may help you. When you have finished (do not try for more than 30 minutes) compare your scheme with that on p. 45.

You will have noted that so far the evidence for a repressor and its mode of action (points 1–8) is purely circumstantial. However, in the past five years, more direct evidence, again from the lactose system, has been obtained.

QUESTION A protein has been isolated from *E. coli* with the characteristics of the postulated repressor. Can you guess what these characteristics are?

ANSWER The protein binds *in vitro* to DNA isolated from the operator region of the lactose operon of *E. coli*. This binding is prevented by the inducer. No such protein can be isolated from constitutive mutants of *E. coli*, showing that this protein is made only by cells with a normal *i* gene and is hence the repressor.

It has recently been possible to mix *E. coli* DNA, RNA polymerase, ribosomes, tRNAs, activating and transfer enzymes, and all the small molecule substrates required, and synthesize β-galactosidase *in vitro*. This synthesis is inhibited by the isolated protein described above. Thus it is almost certainly the repressor.

It should by now be evident that the Jacob-Monod hypothesis, as applied to the lactose operon, fits the observable facts (points 1–8) and is probably correct.

By inference, it is also probably correct for many other cases of induction and repression in bacteria. This suggests that of our alternative hypotheses, hypothesis (2)—that dealing with control of transcription—is probably correct for bacteria. This does not rule out the possibility that some control of translation also occurs. It does, however, seem to be less likely in view of the unstable nature of mRNA in bacteria; i.e. since the mRNA does not survive long it is unnecessary to regulate its utilization (translation) to control the rate of polypeptide synthesis.

Two other points about induction and repression in bacteria should be noted:
(a) When considering repression in branched pathways, an analogous problem to that involved in such pathways in feedback inhibition (Section 5.2.4) occurs. The ways of overcoming it are also analogous, e.g. both end-products are needed to give full repression.

(b) Many enzymes in bacteria are always needed irrespective of environmental conditions. These are enzymes in central pathways such as glycolysis. As one might expect, such enzymes do not seem to be subject to induction and repression, but are made at a steady rate geared to overall growth rate (i.e. constitutively).

Finally, you should appreciate that in many pathways in bacteria both end-product repression and feedback inhibition occur. These two systems, though operating via different mechanisms are, of course, connected as they both have the same response (in terms of metabolic effect) to excess end-product. Their harmonious operation has presumably been achieved over many millions of years of evolution. To give you an example of the complexity of such dual control, we suggest that you look at Figure 18–3 on p. 488 of *Cell Structure and Function*. Do *not* of course try to remember it! (Though, in fact, this is a much simplified version of the real pathway and its control.)

6.4.2 Summary of Section 6.4

This has been a long and somewhat complex Section. We suggest that you try to summarize the major points for yourself. Jot them down, and then check them against our summary given below. If your summary departs substantially from ours, you should look again at Section 6.4.

1 Induction and repression in bacteria involve the synthesis of new polypeptide chains, not just folding of preformed ones.

2 Control of polypeptide synthesis could be due to control of translation of mRNA, or control of transcription.

3 Bacterial mRNAs are unstable.

4 The lactose system gave rise to the Jacob-Monod hypothesis.

5 The Jacob-Monod hypothesis states:

(a) Control is via control of transcription.

(b) Inducible and repressible systems involve operons, i.e. genes for metabolically related enzymes being consecutive on the DNA.

(c) Specific regulator genes exist for each operon.

(d) Each regulator gene contains information for a repressor.

(e) Each repressor is specific for a specific inducer or co-repressor and a specific operator region.

(f) Repressor inhibits mRNA transcription by RNA polymerase by binding to the operator region. This binding of repressor prevents the RNA polymerase moving on to transcribe the structural genes. The binding of repressor is antagonized by inducer in inducible operons, and helped by co-repressor in repressible operons.

6 Some enzymes are made constitutively.

SAQs for Section 6.4

SAQ 9 (*Objective 5*) On adding substance X to some *E. coli* growing on glucose an increase is found in the rate of metabolism of substance X. X is metabolized by enzyme, X-ase.

If the experiment is repeated, but some actinomycin D (a substance that inhibits RNA synthesis) is added along with X, no increase occurs in the rate of metabolism of X. This suggests that X acts by:

1 Stabilizing X-ase against degradation.
2 Stimulating translation of mRNA for X-ase.
3 Stimulating transcription of mRNA for X-ase.
4 Feedback inhibition of X-ase.

Choose the most plausible alternative.

SAQ 10 (*Objective 5*) On adding substance A to some bacteria, two enzymes increase in level: M-ase and N-ase. Both enzymes can use A as a substrate. If puromycin, a substance that inhibits the joining together of amino acids on ribosomes, is added along with A, neither enzyme increases in level. If actinomycin D (an inhibitor of RNA synthesis) is added along with A, only M-ase increases in level. Based on these findings, indicate which of the following statements are likely to be true and which are likely to be false:

1 Both enzymes are synthesized from amino acids in response to A.
2 The information for making both enzymes is on the same mRNA molecule, the genes for the two enzymes being adjacent on the DNA.
3 The regulation of N-ase is completely effected by controlling the translation of its mRNA.
4 The regulation of both N-ase and M-ase is completely effected by controlling the transcription of their mRNAs.

You should now check your answers against those given on p. 46.

6.5 The Regulation of Protein Synthesis in Higher Organisms

In many higher organisms, phenomena similar, at least superficially, to induction and repression in bacteria have been observed. The environment of any one cell in a multicellular organism is obviously more stable than that surrounding a single-celled bacterium. Nevertheless, fluctuations in the levels of various compounds reaching these cells do occur. For example, the level of glucose or amino acids in the blood will vary, depending on how recently the animal has eaten and on the levels of certain hormones (S100, Unit 18). Administration of certain small molecules by feeding or injection can be shown to result in changes in the levels of some proteins in some tissues. Similarly the administration of certain hormones can result in the induction of enzymes in the cells of the target tissues (i.e. those tissues that are affected by the hormones in question). The question therefore arises as to whether these effects seen in higher organisms are due to effects like those in bacteria, involving Jacob-Monod type control systems. This is a very tricky question to answer. Higher organisms, being much more complex than bacteria, are less amenable to experimental study. For example, the site of administration of a compound by feeding is often far removed from its site of action, say the liver. This makes it difficult to conclude that the compound administered is that which is directly responsible for the effect observed; there may be many steps in between. Often the effects on enzyme levels in higher organisms take hours to manifest themselves, not minutes as in bacteria; this again brings into question how directly responsible any experimentally administered compound is for the observed effects. Aside from these experimental difficulties, there are further complications.

1 Some so-called induction in higher organisms is really a very gross phenomenon. For example, certain hormones such as thyroxine (S100, Unit 18), cause big changes in the rate of protein synthesis of hundreds of proteins in the liver. If you look back at Figure 4 (p. 29), you will find that such non-specific effects might be expected if any of the non-specific parts of the system (i.e. parts needed for the synthesis of all proteins) were increased. Indeed it can be shown that thyroxine causes a synthesis of ribosomes, thus allowing a higher rate of all protein synthesis.

We shall not concern ourselves with these non-specific effects in this Unit, fascinating as they are, but limit our discussion to induction more like the specific bacterial variety, where one or a few enzymes are affected.

2 In the cells of higher organisms, unlike bacterial cells, protein degradation is widespread and relatively rapid. This complicates the investigation of enzyme induction and repression, because, as we argued in Section 6.2, induction *could* be due to a decrease in the rate of degradation of the proteins seen to be induced, and repression due to increased degradation.

With these points in mind, it should be clear to you that the current understanding of enzyme induction and repression in higher organisms is, to say the least, complicated and incomplete. Despite the complexities of the experimental systems as compared with bacteria, and despite the points raised in (1) and (2) many investigators have used the Jacob-Monod hypothesis as a basis for explaining induction and repression in higher organisms. It may well be that some such changes in enzyme levels are due to Jacob-Monod-type control of transcription. However, great difficulties arise when trying to explain all the observed data by this hypothesis. We shall now very briefly review some of these difficulties and indicate some other possible explanations for some data.

6.5.1 A critical look at the applicability of the Jacob-Monod hypothesis to higher organisms

The three crucial features of the Jacob-Monod hypothesis as applied to bacteria are as follows:

(a) Specific protein synthesis depends on the level of specific unstable mRNA. Turn off mRNA synthesis, existing mRNA is soon destroyed and the synthesis of the protein ceases.
(b) Inhibition of mRNA synthesis is achieved by the binding of specific repressors to specific operator regions of the DNA.
(c) Repressors are proteins coded for by regulator genes.

The current status of these three features in higher organisms is briefly as follows:

(a) In many cases where enzyme induction is caused by hormones or other agents, it is found that, prior to the observed increase in enzyme, there is an increase in RNA synthesis. In some cases, induction is prevented by the administration of an inhibitor of RNA synthesis (e.g. actinomycin D). These two pieces of data have on occasion been taken to support the contention that the observed inductions are due to increased transcription of specific mRNAs. However, there is often little evidence that the RNA synthesized is solely a specific mRNA rather than due to a general increase in RNA synthesis. Furthermore, studies on the half-lives of mRNAs in higher organisms have revealed that, unlike in bacteria, these mRNAs are of highly varied duration (i.e. the mRNA for one enzyme has a very different half-life from that for another), from a few minutes to several hours or even days. This presents a paradox—if the Jacob-Monod hypothesis is valid for all enzyme induction and repression in higher organisms, how can it operate where the mRNA has a long half-life, since switching-off mRNA synthesis (on removal of the inducer or addition of an end-product) will not result in a rapid cessation of protein synthesis, the stable mRNA still being translated? The synthesis of haemoglobin in reticulocyte cells, mentioned in *Cell Structure and Function* (p. 387), occurs using stable mRNA.

This paradox has led some biochemists to suggest that *some* induction and repression in higher organisms may operate by controlling the rate of translation of these stable mRNAs *or* by controlling their rate of destruction. These controls could in principle operate at one or more of several points, such as the attachment of the mRNA to the ribosomes or the rate of movement of the ribosomes along an mRNA that is being translated. One interesting possible site of control, relevant to the cells of higher organisms only, is due to the existence of a nucleus. Since mRNA is made in the nucleus but translated on ribosomes in the cytoplasm, regulation of protein synthesis could operate by controlling the rate of transfer of mRNA from nucleus to cytoplasm. This suggestion is, in line with many others currently being investigated, somewhat speculative.

(b) and (c) It suffices to say that, though many candidates for the role of repressors and regulator genes (in the Jacob-Monod sense) in higher organisms have been proposed, there is little or no direct evidence to support these contentions. This is not to say that such molecules do not exist in higher organisms, but just that the complexity of the systems studied makes the proof of their existence elusive.

In conclusion, you can see that the current state of understanding of the specific regulation of protein synthesis in higher organisms is highly confused. In some instances, mechanisms of the Jacob-Monod type, that is depending on specific inhibition of mRNA synthesis, probably do operate. In others, regulation is probably at the level of translation. In no one instance can any mechanism be stated with certainty. The only safe conclusion is that any conclusion that one draws at the present time is inconclusive! The situation is further complicated when one recalls that control of protein degradation (Section 6.2) may also be a means of regulating enzyme levels in higher organisms. The longer lifetimes of the cells in higher organisms allow regulation to depend on stable mRNA and, in some instances, relatively unstable proteins. There is good evidence that in some cases, induction is due to stabilization of existing enzymes. This, of course, then raises the question of how this operates, and so on. . . .

One could continue this discussion over many Course Units or even Courses! By now, you will have appreciated the great complexity of this subject. To end this topic, we now suggest that you consider, in the form of an exercise, some data gathered on the induction of one enzyme from a higher organism—tryptophan pyrrolase (TP) in rat liver. For each set of data, we present a set of interpretations. Select at each stage the interpretation that you consider the most justified *based on all the data given up to that point* (i.e. your evaluation should be progressive). Then move to the next set of data, and so on. When you have finished, check your answers against those in the commentary given on p. 39.

If you are short of time, you can save time by reading the commentary immediately after each set of data and interpretations.

The induction of tryptophan pyrrolase (TP)

Data I

A rat was killed, the liver removed and homogenized. The homogenate was then assayed (by an enzyme activity assay) for its content of tryptophan pyrrolase (TP) and protein, thus allowing a calculation of the specific activity (enzyme activity per g protein) in the homogenate. Some other rats were divided into 4 groups and treated as follows:

Group 1   Each rat was injected with the hormone hydrocortisone (HC).
Group 2   Each rat was injected with tryptophan (try).
Group 3   Each rat was injected with HC and try.
Group 4   Each rat was injected with a salt solution. This was the control group.

Every four hours, one rat from each of the four groups was killed, the livers removed, homogenized and assayed for TP and protein. The specific activities of TP for rats in the four groups are plotted versus time in Figure 8.

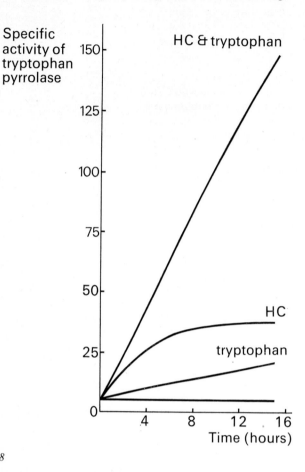

Figure 8

Interpretations I

A HC and try given independently or together increase the synthesis of mRNA for TP.

B HC and try increase the level of TP by different mechanisms.

C HC and try increase the level of TP by similar mechanisms.

D HC inhibits the induction of TP by try.

Data II

Some more rats were divided into 3 groups and treated as follows:

Group 1 The control group. Each rat was injected with a salt solution.
Group 2 Each rat was injected with HC.
Group 3 Each rat was injected with try.

Three hours later all the rats were injected with radioactive amino acids. Some hours later all the rats were killed, the livers removed, homogenized and some of the homogenate in each case assayed for TP and protein, and hence specific enzyme activity. TP was purified from the remaining portions of each of the homogenates and the specific radioactivity of the purified TP samples measured. The results are shown below. The specific activity of TP (i.e. enzyme activity per g protein) is in the left-hand column, the specific radioactivity of the TP (counts per minute $\times 10^{-3}$ g purified TP) in the right-hand column.

Group	Specific activity of TP	Specific radioactivity of TP
1 (salt)	40	1400
2 (HC)	160	9500
3 (try)	80	1900

These figures are approximate and no special significance should be attached to their exact values.

Interpretations II

A HC and try both act primarily by increasing the synthesis of TP.
B HC acts primarily by increasing the synthesis of TP, try does not.
C HC and try act by similar mechanisms to increase the level of TP.

Data III

Some rats were divided into each of two groups (A and B). Each rat in both groups was injected with radioactive amino acids. Later, each rat in group A was injected with salt solution, each rat in group B with try. At various times later, rats from each group were killed. TP was purified from the livers and its specific radioactivity determined. The results are shown in Figure 9.

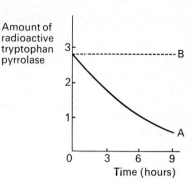

Figure 9

Interpretations III

A Try acts by degrading TP.
B HC acts by causing increased synthesis of TP; try acts by causing reduced degradation of TP.
C Try causes increased synthesis of TP; HC causes reduced degradation.
D HC and try both act primarily by causing increased synthesis of TP.

Commentary

Comments on I

Interpretation B is best. The different time course and shapes of the curves (Fig. 8) for HC and try, and the fact that the two together give a more than additive increase suggest that they affect the level of TP by different mechanisms.

A is therefore wrong. Anyhow no evidence for the involvement of RNA synthesis is presented.

C is wrong.

D is obviously wrong as HC+try (together) produce a greater effect than try alone.

Comments on II

Bear in mind that a radioactive TP sample indicates that some of the enzyme has been synthesized after injection of the radioactive amino acids. The differences between the control group (1) values and those of the other groups, therefore, give some idea of the amount of TP synthesized in response to HC or try. Therefore, one finds that HC causes a much greater stimulation of TP synthesis than does try. This suggests that interpretation B is the correct one.

However, if HC causes a much greater stimulation of TP synthesis than does try, one is left to explain the fact that the final amounts of TP in the homogenate, as measured by the specific enzyme activity (left-hand column and Fig. 8), are not that widely different. This is dealt with in the next Sections.

Comments on III

B is almost certainly correct. Data II point to HC leading to increased TP synthesis. Data III show that try, injected later, stabilizes pre-formed radioactive TP.

A is therefore wrong.

C is therefore wrong.

D is wrong (see I and II).

Therefore, what is probably happening is that TP is synthesized and degraded in rat liver at a low steady rate even in the absence of added HC or try (note that in Fig. 8 TP is never zero). HC causes increased synthesis. Try stabilizes any enzyme present. (But try does not stimulate the rate of TP synthesis.)

The more than additive effect seen in Figure 8 is now also readily explicable. For example, if HC leads to a tenfold increase in the level of TP owing to increased synthesis, and try renders this five times more stable, then the net result when both HC and try are present will be a $10 \times 5 = 50$-fold increase in TP.

6.6 The Control of DNA Synthesis and Cell Division

We have now completed our discussion of the regulation of cell metabolism by control of the activity (Unit 5) and the amounts of enzymes. However, as we pointed out in Section 5.1, in order to survive successfully, an organism must also have a means of regulating DNA synthesis and cell division. Any one cell contains a complete set of genetic information in the form of DNA. At some time during the cell cycle (i.e. from one division to the next), this DNA must be replicated to give two sets of genetic information. These two sets must be separated and then partitioned equally amongst the two daughter cells. Failure to replicate the DNA exactly or partition it accurately in each cycle results in two daughter cells which differ. So the whole basis of genetic continuity depends on the accuracy of regulating these processes.

As you know from S100, Unit 17, the basic chemistry of DNA indicates the means by which replication can occur, that is by utilizing base complementarity. That this actually occurs is borne out by the elegant experiments of Meselson and Stahl (S100, Unit 17). However, how this process is regulated is not well understood. Regulation does indeed occur. In bacterial cells, the rate of DNA synthesis is linked to the rate of overall cell metabolism. Cells metabolizing rapidly, as when in a medium rich in foodstuffs, synthesize DNA more rapidly and divide more often than cells in a medium poor in foodstuffs. In cells of higher organisms, DNA synthesis is known to occur at strictly controlled times in the cell cycle, and such cells have evolved a very precise mechanism for partitioning one set of genetic information into each of two daughter cells: the whole finely organized process of mitosis (S100, Unit 17).

Once again, more is known about the details of these processes of DNA synthesis and cell division in bacteria than in higher organisms, but even in bacteria the questions far outnumber the answers. We cannot even begin to discuss here all the details or indicate the controversies—of which there are all too many—so we shall just mention a few of the fairly well established features of the regulation of DNA synthesis in bacteria that have been studied, and then go on very briefly to consider higher organisms.

(a) Bacteria

In the bacteria so far studied all the genetic information appears to be contained in a single long molecule of DNA—one long double helix. The process of replicating this molecule can be subdivided into 3 steps: (i) initiation; (ii) chain growth; (iii) termination.

Initiation is the process wherein DNA replication starts. It is thought that, for any one molecule of DNA, initiation occurs only at one fixed point, an *initiator*. This is presumably a special sequence of nucleotide bases. It is not definitely known what triggers initiation, but probably it is the interaction of some 'initiator substance' with the initiator.

Following initiation, *chain growth* commences. This involves a progressive separation of the strands of the double helix. The enzyme responsible for copying the sequence of nucleotides, DNA polymerase, attaches to the initiator and starts joining nucleotide base to nucleotide base to produce new polynucleotide chains. When the whole molecule of DNA is completely copied to give two molecules, the process is completed—*termination*.

These events are summarized in Figure 10.

In principle, either initiation or chain growth are points at which the rate of DNA synthesis could be regulated. Thus the frequency at which initiation takes place or the rate at which DNA polymerase moves along the DNA adding nucleotides together could be altered. Recent work on *E. coli* has suggested that it is the frequency of initiation that is the crucial point of control, though in very slowly growing cells the rate of chain growth may also be slowed. So the rate at which a bacterial cell is synthesizing macromolecules such as proteins, which depends on the richness of the medium, governs the frequency of initiation of DNA synthesis, perhaps, by determining the level of an initiator substance. Further studies have implied that the processes culminating in cell division commence following termination. Thus the rate of overall protein synthesis due to the richness of the medium regulates the frequency of DNA synthesis which in turn regulates cell division.

Figure 10 Replication of DNA. (a) The enzyme, DNA polymerase, is about to attach to the initiator. (b) Chain growth is occurring, leading to a 'replicating fork' (i.e. the forked structure indicated). The 'new' chains are shown as red dotted lines. The direction of replication is from left to right. (c) The DNA gas been completely copied to give two new molecules. The DNA polymerase has detached.

The contents of the above paragraph are still highly speculative, but they do at least provide a reasonable working hypothesis linking DNA synthesis, cell growth, and cell division in bacteria. But we have still to discuss the partition of the DNA into the two daughter cells. In essence the problem is this—if a cell contains two molecules of DNA, how does it ensure that the DNA is partitioned 1:1 rather than 2:0? The answer seems to lie in the fact that DNA appears to be attached to the cell membrane, so, as it is replicated, it is segregated to give two molecules attached to the membrane. If a cell membrane dividing off the cell into two cells now forms between the two attached DNA molecules, the problem is solved. (However, the *if* is a big one.) This process, much schematized, is represented in Figure 11. It may well be that the DNA polymerase is at the point of attachment in the membrane and the DNA travels past the DNA polymerase during replication rather than the other way round. The net result would be the same.

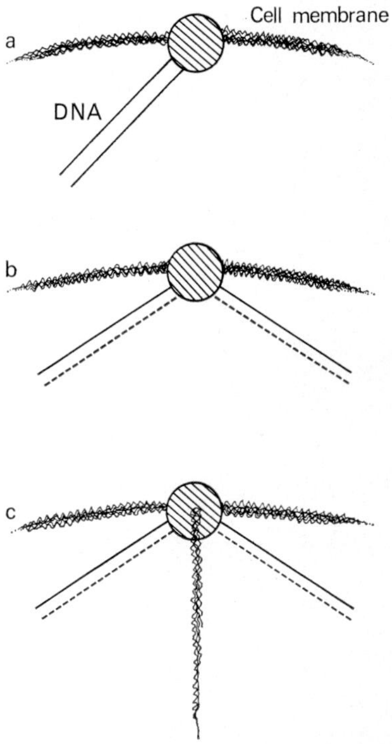

(b) Higher organisms

In higher organisms, the DNA is contained in the nucleus with proteins to form structures called chromosomes, which are clearly visible under the microscope during mitosis. Since it is not even known whether one chromosome contains one or more molecules of DNA, the processes governing DNA replication in higher organisms are barely understood at all. One knows how chromosomes are partitioned between two cells by mitosis, but how is the DNA once synthesized partitioned into each of two new chromosomes? Some data is now accumulating on these interesting and important questions, but as everything is so speculative we will not discuss them here. For those of you who have survived the intricacies so far, and who wish to examine the controversies in detail, we recommend the third level course on *Biochemistry and Molecular Biology*.*

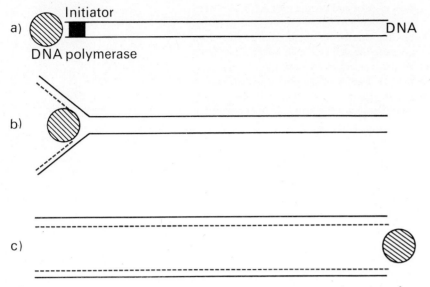

Figure 11 (a) One molecule of DNA attached to the cell membrane. (b) Two 'new' molecules of DNA attached to the cell membrane. (c) New cell membrane partitioning the DNA molecules.

6.7 Conclusions to Units 5–6

In these two Units, we have ranged far and wide over the field of cellular regulation. You are by now no doubt aware of the complexities of this field. You can also see that what is true for *E. coli* is not necessarily true for *E. lephant*, or for the other organisms in between. However, with the guide-lines provided by the extensive work done on *E. coli* and other bacteria, it should be possible to tackle effectively the elephantine problems posed by higher organisms.

We must, however, conclude with one final note of caution. In choosing our examples or in the experiments we have described, one is always dealing with

*This Course will be introduced in 1976.

extreme situations. For example, large excesses of end-products or substrates were added experimentally to make the effects great enough to observe and describe. However, in the living cell, presumably, the changes in the internal and external environments are rarely so dramatic. Hence the cell's control systems probably operate in a mildly fluctuating, on-off manner, so as to keep the internal conditions in the cell at a relatively steady level. This principle of course also applies to regulation of whole cells, tissues and organs, in whole organisms. It is a principle with which you are by now very familiar—the principle of *homeostasis*.

Conclusion to the Course

The six short Units of this Course have taken you at a fair gallop through a number of topics which are of importance to contemporary biochemistry. As we said in the *Introduction and Guide*, we have chosen to concentrate on a small number of main themes; the relationship between structure and function in biological macromolecules; the regulation and control of enzymes and the metabolic pathways in which they are concerned; the problem of cell energetics and the roles of ATP. These themes are not so different from those of many biochemistry courses at conventional universities; where the difference does lie is that, in this short Course, we have tried to eliminate redundant material and use examples only for illustration. One other major theme has been, of course, the experimental techniques of biochemistry; you have seen them demonstrated in the TV programmes, and by now should have some experience of them through your own Home Experiment.

For those of you who will do no more biochemistry than this, we hope the Course has at least made you familiar with some of the key themes, jargon and methodology of the science. Those of you who combine *Biochemistry* with such parallel second level courses as that on *Genetics and Development** will see how certain concepts overlap, linking together these two central areas of today's biology. But those who wish to deal more rigorously and in greater depth with issues that we have had to pass over very briefly here, will need to come back to biochemistry at third level, to the half-credit course on *Biochemistry and Molecular Biology*.

Meanwhile, we hope we have convinced you of biochemistry's claim to illumine and unify many diffuse areas of biology. But biochemistry is not just research about theoretical questions, using complex and exotic equipment. Here and there in these Units, and more systematically in the final TV programme, we have referred to the interactions of biochemical research with industrialized society. In a sense, the food industry, agriculture and medicine have always made use of biochemical procedure, albeit often derived by trial and error, without supporting theory. Today, advances in all three are most likely to be made only by the linking of theory with practice: biochemical research labs will be found associated with hospitals, cosmetic firms and the oil industry, among many others. Clearly, as the success of biochemical reductionism in explaining more and more of biology continues in years to come, so the role of biochemists in directly applying their skills in industry and society at large will also expand.

Not all these applications are desirable or have beneficial consequences, of course. Thalidomide and numerous potential pollutants—to say nothing of drugs for social control and chemicals for warfare—are part of the harvest of biochemistry. Sagas such as those on the enzyme washing powders, the organophosphorus flykillers and the vaginal deodorants—which introduce new and potentially dangerous substances into the environment of the factory worker who makes them or of the consumer who uses them—may indicate some of the consequences of the rash utilization of biochemical knowledge. But as you learned from S100, the development of scientific knowledge is not just a one-way affair, with biochemical advances occurring in a separate chamber from which they trickle out at random to be applied socially. Rather we are dealing with mutual interactions: social demands are made on biochemists, and these alter the direction of their work and the way they see the major problems of their field. These interactions are both theoretical and practical in nature. The central concepts we have adopted in this Course—'energy', 'control', 'structure and function' are not mere abstractions. They are concepts that have meaning in the scientist's everyday life, and his everyday experience provides the conceptual 'set' for his sciences. Whether we analyse the cell in terms of energy exchanges, with ATP as a central theme, or as a sequence of flows of information, with regulation and control mechanisms as central themes, we are in either case adopting a standpoint in relationship to our subject-matter which cannot be isolated from the social environment in which our research is conducted. If we were to see society in a different way, would we see our biochemistry differently also? We can ask these questions more easily than we can answer them; we leave you to think about them.

**This Course will be introduced in 1973.*

Answers to Pre-Unit Assessment Test

Section A

1 That catalysed by M-ase.
2 D.
3 True.
4 True.
5 False, e.g. control of heart rate (S100, Unit 18).
6 True.
7 False. The science of cybernetics demonstrates the similarity (S100, Unit 16).

If you were wrong on more than two of these questions, you should read again pp. 187–201 of *The Chemistry of Life* before proceeding with this Unit.

Section B

1 False. RNA contains ribose, but DNA contains deoxyribose.
2 True.
3 False. For each of the 20 amino acids, except tryptophan and methionine, there are two or more codons.
4 False.
5 True.
6 True.
7 False. Transcription occurs on DNA.

If you were wrong on more than two of these questions, you should read quickly through S100, Unit 17, paying particular attention to the Sections relevant to the questions on which you were wrong, before proceeding with this Unit.

Section C

Only in questions 1, 2 and 3 are the statements teleological.

If you identified any of the statements incorrectly, you should read again the Section on teleology in S100, Unit 18, before proceeding with this Unit.

Answers to In-text Questions

ITQ 1 (*Objective 3*)   The correct order of the pathway is:

$$P \xrightarrow{f} R \xrightarrow{a} T \xrightarrow{e} Q \xrightarrow{b} V \xrightarrow{c} W \xrightarrow{d} Y$$

The mutants a–f are blocked (i.e. lack an enzyme) at the steps indicated. There are many ways of solving this problem, but consider the one below. Since there are five known intermediates (R, Q, V, W and T) plus P and Y, there must be at least six steps. Mutant d will not grow on P plus any of the intermediates, therefore it must be blocked after the intermediates, so that it cannot convert any of them to Y. In other words, it is blocked in the last step. Since d accumulates W, this must occur just before Y, hence

$$P \longrightarrow ? \longrightarrow ? \longrightarrow ? \longrightarrow ? \xrightarrow{d} W \longrightarrow Y$$

Mutant c can grow only on P plus W. It must therefore be blocked before W but after the other mutants. Hence the compound it accumulates, V, must occur just before W, hence:

$$P \longrightarrow ? \longrightarrow ? \longrightarrow ? \xrightarrow{c} V \xrightarrow{d} W \longrightarrow Y$$

and so on.

If you still find this reasoning difficult, look at the pathway given above, treat each step in turn and then imagine the consequences to the organism if that step were blocked, i.e. what would it accumulate from P and from which compounds could it produce Y and hence grow.

ITQ 2 (*Objective 4*)

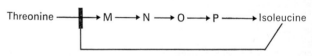

If you got this wrong, check back to p. 16.

ITQ 3 (*Objectives 5, 7 and 9*)   From Figure 4 it is evident that the only components necessary for the scheme shown in Figure 5 are completed polypeptides. Therefore, all the other components leading up to these (ATP, GTP, mRNA, DNA, amino acids, etc.) are unnecessary. Therefore, the scheme in Figure 5 could be tested by two types of experiment:

1 One could add inhibitors of any of the steps prior to completion of polypeptides and see if these inhibited induction.

2 One could add radioactive amino acids during induction, isolate the induced enzyme and see if it was radioactive.

In essence, this is what statements (a) and (b) relate to. If the scheme in Figure 5 were true then (a) would be true and (b) would not be true, as amino acids and peptide bond formation are irrelevant to the stage where polypeptides are folded to give active enzyme.

ITQ 4 (*Objectives 8 and 9*) Repression can be explained by the Jacob–Monod hypothesis as follows. A regulator gene exists for each operon repressed. This regulator gene contains the information for a protein repressor. This repressor, on binding to the operator region, inhibits synthesis of mRNA from the structural genes of the operon in question. However, the repressor can only bind to the operator if it first combines with a small molecule called the co-repressor. The co-repressor is the end-product or some substance derived from it. So the sequence of events for repression by tryptophan are as follows:

1. Tryptophan enters the cells where it is converted to co-repressor.
2. Co-repressor binds to repressor.
3. Repressor (+co-repressor) now binds to the operator.
4. Further mRNA synthesis soon ceases as RNA polymerase now cannot pass the operator.
5. Existing mRNA is destroyed.
6. Synthesis of the enzymes ceases.

These events are summarized in Figure 12.

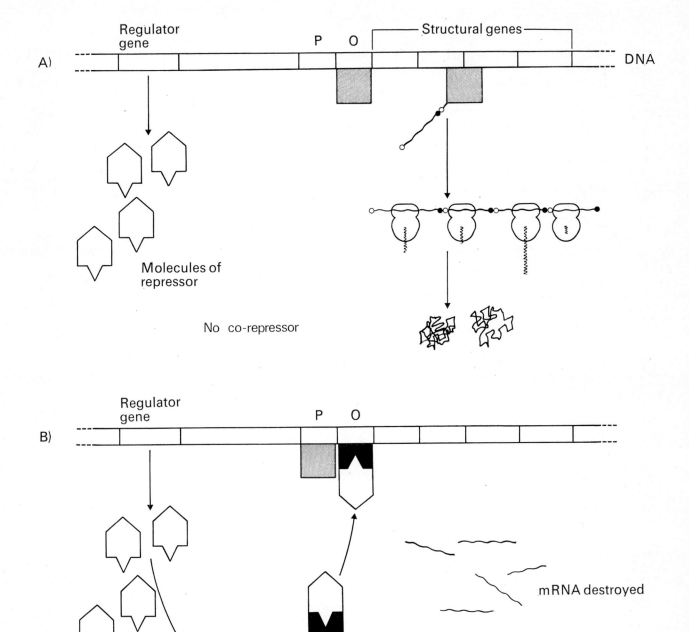

Figure 12 RNA polymerase is shown in red. In (A), transcription is occurring; two molecules of RNA polymerase are shown here, one just having reached the operator region (O), the other having by now travelled halfway along the structural genes. In (B), transcription has ceased owing to repressor + co-repressor having bound to the operator region.

45

Self-assessment Answers and Comments

SAQ 1 (*Objective 4*) (a) It will inhibit Steps 5 and 1 (via 1h). Assuming the production of B by 1h is just enough to produce the amount required for the pathway leading to H, then only the synthesis of B, C, D, G and H will decrease. Hence B, C, D and G will be decreased in amount.

(b) D will give rise to E, F, G and H. Therefore a large amount of D will lead to an excess of F and H. This will lead to inhibition of Steps 1 (1f and 1h) and 4 and 5. Therefore, synthesis of all the compounds from A will be reduced.

(c) Excess F will inhibit Step 4 but not Step 1 (as 1f is missing). Therefore, normal production (that is normal for this mutant lacking 1f) of B, C and D will ensue. However, all of D will now go to produce H. Consequently, excess H will build up. This will then inhibit Steps 5 and 1 (via 1h). So the whole pathway will be eventually inhibited.

SAQ 2 *Objectives (1 and 5)*
(a) False (Section 5.3).
(b) True (Section 5.3).
(c) True (Section 5.3).
(d) False (Section 5.4)

SAQ 3 (*Objective 5*)
(a) True (Section 5.3).
(b) False (Section 5.3).

If you were wrong on any parts of these questions, you should look back at the relevant Sections.

SAQ 4 (*Objectives 1 and 5*)
(a) (iii) is the best alternative.
 (i) is wrong as repression would lead to *decreased* enzyme.
 (ii) is wrong as feedback inhibition would not affect the amount of histidase present, just *inhibit* its activity.
 (iv) is wrong as this would reduce the activity of the enzyme not increase its amount.

(b) (i) is correct.
 (ii) is wrong (Section 5.2.3).
 (iii) is wrong (Unit 2).
 (iv) is wrong (Unit 2).

SAQ 5 (*Objectives 3 and 5*)
(a) (i) or (iii) could be correct. Radioactive arginine indicates arginine synthesis by the cells. (i) however, is most likely, as it alone, acting rapidly, could result in the cessation of arginine synthesis.

(b) (ii) is the best alternative, since if (i) were correct puromycin should inhibit the renewed synthesis of arginine; similar arguments rule out (iii).

If you were wrong on any parts of these questions you should look back at the relevant Sections.

SAQ 6 (*Objective 7*) (a), (c), (d), (f), (g).

SAQ 7 (*Objective 6*)
(a) False. They are ribosomes attached to pieces of membrane.
(b) False. mRNA does.
(c) True.
(d) False. They also contain protein.
(e) True.
(f) False. It is universal.

SAQ 8 (*Objective 6*) (a) Since ribosomes are strung together in a polysome along mRNA, breaking this RNA with ribonuclease disrupts the polysome to yield single ribosomes.

SAQ 9 (*Objective 5*) 3 is the most plausible alternative since inhibition of RNA synthesis by actinomycin D blocks the increase in X-ase (Fig. 4).

SAQ 10 (*Objective 5*)
1 True. Since puromycin inhibits the increase in both.

2 False. Since this would mean that actinomycin D should inhibit the increase in both or none as both genes would be transcribed at the same time.

3 False. Since actinomycin D (an inhibitor of transcription) inhibits the increase in N-ase.

4 False. Since actinomycin D does not inhibit the increase in M-ase.

BIOCHEMISTRY

1 Biological Macromolecules
2 Enzymes
3 Cell Energetics, the Mitochondrion and the Chloroplast
4 Metabolic Pathways
5⎫
6⎭ Regulation of Cell Processes